CRUSH YOUR MATH FEAR!

TIPS, TRICKS, & RIDDLES TO IMPROVE YOUR MATH SKILLS

David Frush

KidsSci Publications
KidsSci Inc.
Chicago, IL

Copyright © 2017 KidSci Inc.

All rights reserved. No part of this publication may be reproduced, distributed, or transmitted in any form or by any means, including photocopying, recording, or other electronic or mechanical methods, without the prior written permission of the publisher, except in the case of brief quotations embodied in critical reviews and certain other noncommercial uses permitted by copyright law. For permission requests, write to the publisher, addressed "Attention: Permissions Coordinator," at the address below.

KidSci Publications
A Division of KidsSci Inc.
9445 W Majestic Drive
Monee, IL 60449-7111
https://KidsSci.com

Ordering Information:

Special discounts are available on quantity purchases by corporations, associations, and others. For details contact the publisher at the address above.

Orders by U.S. trade bookstores and wholesalers. Please contact KidsSci Publications: Tel: (815) 534-1111 or email Publications@KidsSci.com.

Printed in the United States of America

Library of Congress Control Number: 2017951878
Frush, David C.
Crush Your Math Fear! Tips, Tricks, & Riddles to Improve Your Math Skills

ISBN- 978-0-9992444-1-8
ISBN-10: 0999244418

First Edition

17 16 15 14 13 12 11 10 9 8 7 6 5 4 3 2 1

For Marliesa

Table of Contents

MATH TIPS 1

    LEFT HAND ADDITION 2

    QUICKLY ADD NUMBERS OF 3 OR MORE DIGITS 4

    YET ANOTHER WAY TO ADD COLUMNS OF NUMBERS 5

    DOT ADDITION 8

    A QUICKER WAY TO ADD A COLUMN WITH MULTIPLE DIGITS 11

    MORE LEFT HAND ADDITION 13

    EASIER SUBTRACTION BY ROUNDING 15

    SUBTRACTION BY INSPECTION 17

    HOW TO MULTIPLY ANY NUMBER BY 11 19

    QUICKLY MULTIPLY NUMBERS ENDING IN 5 22

    QUICKLY SQUARE NUMBERS ENDING IN 5 24

    A QUICK WAY TO MULTIPLY WHEN YOUR MULTIPLIER ENDS IN 1 .. 25

    A QUICK WAY TO MULTIPLY WHEN THE MULTIPLIER IS A COMPOSITE NUMBER 28

    A QUICK WAY TO MULTIPLY BY 15, 150, & 1,500 30

    HOW TO MULTIPLY BY COMPLEMENTS 32

    HOW TO MULTIPLY BY EXCESSES 36

    QUICKLY MULTIPLY ALTERNATING NUMBERS 38

    QUICKLY SQUARE ANY NUMBER OF TWO DIGITS 39

    HOW TO FIND THE SUM OF ANY SEQUENCE OF NUMBERS 42

How to Calculate the Sum of a Geometric Series 47

Division Tips 49

Fraction Tips 54

How to Find Square Roots Without a Calculator 61

Another Way to Find Square Roots......................................63

You Know More Latin Than You Think! 65

Prime Numbers Tips 66

MATH TRICKS...67

The 1,089 Math Trick 68

The Number Thought Upon Trick 72

The Fibonacci Trick 74

The 99 Math Trick 80

The Guess the Number Trick.. 82

The Instant Addition Trick.. 83

The Dice Trick..85

The Day of the Week Trick 87

Another Day of the Week Trick.. 91

Another Number Thought Upon Trick 93

The Number 34 Trick 95

The Addition Table Trick.. 97

Yet Another Number Thought Upon Trick 100

The Multiplication Summation Trick 102

- The 1969 Trick 105
- The Tower of Hanoi Trick 107
- A Mindreading Trick 110
- The Birth Month & Year Trick 113
- The Lucky Numbers Trick 115
- A Squaring Trick 116
- The Sequential Remainders Trick 117
- Numerical Tricks 119
- The Number 9 Trick 122
- The Card Whisperer Trick 124
- The Gergonne Card Trick 128
- Howard Adams Mated Cards Trick 131
- A Classroom Card Trick 136
- The Magic Age Table Trick 138

MATH RIDDLES 141

- Math Riddle Solutions 165

ABOUT THE AUTHOR 227

Preface

I am always looking for fun and entertaining ways to teach kids math and science and this book started as a journal of things I found interesting in some very old math books. Believe it or not, hundreds of years ago and up to the first half of the last century people used math as a form of entertainment. Many of the math tricks in this book date back to as far as the late 1600's and one of the riddles is based on one created around 250 AD. The math tips are from a time long before the calculator was invented when people routinely had to do mental arithmetic. I've tweaked these tips, tricks, and riddles a bit, made them more readable for today's readers, and provided complete explanations for how to solve everything.

Introduction

A recent survey found 53% of 18- to 35-year-olds think they aren't good at math. Saying you're not good at math is like saying you're bad at playing the violin when you haven't touched one in years. Math is a skill, like playing a musical instrument or sports, and if you don't practice and make those neural connections in the math part of your brain you're never going to improve. Math isn't easy and like most things not easy people go out of their way to avoid it. My hope is this book will encourage you to practice your math skills and provide you with some entertainment at the same time.

Math Tips

This section will teach you how to do over two dozen math shortcuts, some will teach you mental arithmetic and others are meant to be done on paper. You will find these very useful after the zombie apocalypse or when IBM's Watson becomes self-aware and there are no more calculators or smart phones to use as mental crutches.

I highly recommend you try all the tips because putting pencil to paper and stretching your mind by doing math differently is an excellent way to exercise your brain. This will also benefit you if you are someone who learns better by doing rather than just reading or listening. Follow along with the book and then make up your own problems for additional practice. Once you've learned all the tips try combining them and you'll be surprised to find you can do some complicated math in your head.

Spending time learning and using these tips will improve your math skills and greatly increase your confidence.

Left Hand Addition

Let's say you need to add a lot of numbers and horror of horrors, there isn't a calculator anywhere to be found. No laptop, tablet, smart phone, smart watch, not even an abacus. You're going to have to plow through these numbers manually, and you need the answer fast. Fast like the sum of these numbers is the key code to diffuse a doomsday device set to detonate in 59 seconds, 58, 57....

Would you trust yourself to add all the numbers correctly? What if you didn't have time to double check your answer? One wrong digit and BOOM! There goes the city. Or building. Or cruise ship. Or whatever your imagination comes up with, something a bad guy would blow up with a doomsday device. Right about now how would you like some foolproof ways of adding lots of numbers? We have not one, not two, but six different tips for how you can quickly and easily sum large groups of numbers.

Let's try a method called left hand addition first. To add the column of seven two-digit numbers below start at the bottom of the column and work your way up the list, going from right to left, from the ones place then to the tens place:

```
   24      304 + 4 = 308 + 20 = 328
   36      268 + 6 = 274 + 30 = 304,
   21      247 +1=248 +20 = 268,
   83      164 + 3 = 167 + 80 = 247,
   62      102 + 2 = 104 +60 = 164,
   53      49 + 3 = 52 + 50 = 102,
+  49
  ───             **Start at the bottom**
  328             **& work your way up,**
                   **right to left**
```

This will also work for columns of 3 or more, just keep working your way up from the bottom and go from right to left:

```
   142     1,061 + 2 = 1,063 + 40 = 1,103 + 100 = 1,203
   381     680 + 1 = 681 + 80 = 761 + 300 = 1,061,
   212     468 + 2 = 470 +10 = 480 + 200 = 680,
+  468
 ─────
 1,203
```

You'll find it's much easier to do multiple small addition problems rather than going column by column and remembering to carry a number like you were taught in elementary school. There are also fewer steps involved in doing it this way since we're not carrying numbers to the next column, and you have the added benefit of looking cool when you do it in your head!

Quickly Add Numbers of 3 or More Digits

To add two three-digit numbers sometimes it's easier to break it down into two or three smaller problems. Take this problem for example:

$$\begin{array}{r} 376 \\ + 234 \\ \hline 610 \end{array} \quad \begin{array}{l} \textit{Round down to} \rightarrow 370 \\ \textit{Round down to} \rightarrow + 230 \\ \textit{add in } 6 + 4 = 10 \quad 600 + 10 = 610 \end{array}$$

This also works if you want to round up, you just need to subtract the numbers you used to round up:

$$\begin{array}{r} 376 \\ + 234 \\ \hline 610 \end{array} \quad \begin{array}{l} \textit{Round up to} \rightarrow 380 \\ \textit{Round up to} \rightarrow + 240 \\ \textit{Subtract } 4 + 6 = 10 \quad 620 - 10 = 610 \end{array}$$

This could come in handy when you're shopping and you need to sum dollar amounts. For example,

$47.83	*Is the*	$40	$7	80¢	3¢
+ $38.46	*same*	+ $30	+ $8	+ 40¢	+ 6¢
$86.29	*as*	$70	$15	$1.20	9¢

$85 $86.20 $86.29

If you just need a quick estimate of how much something is going to cost leave off the pennies or the change altogether.

Yet Another Way to Add Columns of Numbers

As the saying I just made up goes there is more than one way to peel an orange (my family would be very upset with me if I used the other saying). This method of adding columns of numbers is particularly useful if you need to add really long columns of one or two digits. Let's try it with the column of eleven numbers in the example below:

```
  8  9              Step 1: Starting from
  4                 the bottom right and
  2                 working your way up
  4                 add consecutive digits
  1                 until you get as close to
  6  7              20 as possible. Write
  6                 the ones digit of the
  5                 sum to the right of
  4  7              where you stopped and
  6                 repeat this process
 +7     7 + 7 + 9 = 23   until you reach the top.
 ──
 53     2 + 3 = 5
```

Step 2: Add up the side numbers and whatever numbers are left at the top and the ones digit of this sum is the ones digit of your answer. In the example on the left $7 + 7 + 9 = 23$ so the ones digit of the answer will be 3.

Step 3: To find the tens digit add the number carried over, the 2 from 23, and add it to the number of digits you have on the side, 3, so $2 + 3 = 5$ and your tens digit is a 5.

This also works with columns of two or more digits, let's try a column of fourteen two-digit numbers:

```
                    9    38
                         42   6
                         12
                         51
                         67
                    8    54
                         85   7
                         55
                    5    57
                         20   4
                         14
                         74
                    7    80
7 + 5 + 8 + 9 = 29      + 46       4 + 7 + 6 + 8 = 25
2 + 4 = 6                695       2 + 3 = 5
```

Step 1: Use the same process as before with the right column. The ones digit from the sum of the side numbers plus what is left at the top is the ones digit of your answer.

Step 2: The tens digit of your answer is the ones digit of the sum of the left side numbers.

Step 3: You find the hundreds digit the same way we found the tens digit in the previous example, we add the 2 we carried to the number of digits on the left, 4, to get 6 for the hundreds digit.

So why would anyone in their right mind want to add a column of numbers this way? It may look like more work but the benefit of this method is a) you're always adding relatively small numbers, mostly under 20, and b) the side numbers keep you from losing your place. It's no fun to get ⅔'s of the way through a long addition problem only to realize you're not 100% sure about the last number. Then there are only two possibilities, assume you didn't lose your place and press on and risk a wrong answer or start the problem all over again. Most people choose the second option. If you're only adding a few numbers at a time and keeping track of your progress on the sides you will never lose your place again, so why not give it a try?

DID YOU KNOW?

The word algebra is a Latin variant of the Arabic word al-jabr. This came from the title of a book, Hidab al-jabr wal-muqubala, written in Baghdad about 825 AD by the Arab mathematician Muḥammad ibn Mūsā al-Khwārizmī. Al-jabr was one of the two operations he used to solve quadratic equations. He is generally recognized as one of the fathers of algebra. The word algorithm also came from al-Khwārizmī, it was derived from Algoritmi, the Latin form of his name.

Dot Addition

This method is like the previous one with one small change. This time when you've added up as close to 20 as possible mark it with a dot and continue adding using the ones digit from the location of the dot. For example, in the column below add 6 + 3 + 6 =15 and mark it there with a dot because adding 7 puts us over 20. We continue adding up using the 5 from the 15, add the next 7 to get 12, and make another dot because adding 8 puts us up to 20. At the top the last dot stops at 15 so our ones digit is a 5, and counting the dots gives us 6 which is the tens digit.

```
  8  •   7 + 8 = 15
  9  •   8 + 9 = 17
  7  •   8 + 3 + 7 = 18
  3
  2  •   2 + 8 + 6 + 2 = 18
  6
  8
  7  •   5 + 7 = 12
  6  •   6 + 3 + 6 = 15
  3
 +6
 ___
 65
```

Step 1: Starting from the bottom add consecutive digits until you get as close to 20 as possible, them make a dot.

Step 2: Starting with the ones digit you just found add consecutive digits until you get close to 20 again and make another dot. Repeat until you get to the top and the ones digit of your last sum is the ones digit of the answer. In the example to the left the last sum was 15 so the ones digit is 5.

Step 3: To find the tens digit count the number of dots, our example has 6 dots so the tens digit is 6.

This also works for columns of two digits, just add up the second column using the number of dots from the ones column as a starting point. In the example below we ended up with 4 dots on the right so we add up from the bottom of the tens column starting with 4. Then the ones digit from the last sum at the top of the tens column becomes the second digit and the number of dots gives you the hundreds digit.

8 + 4 + 3 = 15	•	38	•	8 + 8 = 16
		42	•	6 + 7 + 1 + 2 + 2 = 18
6 + 6 + 5 + 1 = 18	•	12		
		51		
		67		
3 + 8 + 5 = 16	•	54	•	7 + 5 + 4 = 16
		85		
8 + 5 = 13	•	55	•	5 + 7 + 5 = 17
6 + 7 + 5 = 18	•	57		
		78	•	1 + 6 + 8 = 15
5 + 9 + 2 = 16	•	26		
		+ 91		
		656		

Step 1: Starting from the bottom right add up to 20 as close as you can and make a dot. Repeat until you reach the top.

Step 2: Add the number of dots on the right to the bottom left digits until you get close to 20 and make a dot. Repeat to the top and the ones digit of your last sum is the second digit of the answer.

Step 3: The number of dots on the left is the hundreds digit.

I prefer dot addition over the previous method using numbers on the side. I think the dots are neater than using numbers plus anyone looking at your work will wonder what the heck you're doing. However, it's really only useful for one- and two-digit columns of numbers, so if you need to add larger numbers you may want to try one of the following methods.

DID YOU KNOW?

Ancient civilizations knew the ratio of a circle's circumference to its diameter was about 3 but Archimedes (c. 287 - c. 217 BC), a Greek mathematician, physicist, engineer, inventor, and astronomer, was the first to calculate it by using a 96-sided polygon to come up with the ratio 22/7, or 3.14285714, which is pretty darn good considering pi is 3.14159265.

A Quicker Way to Add a Column with Multiple Digits

The previous two methods can quickly become cumbersome with columns of three or more digits. Instead of adding up as close to twenty as you can add the entire column and write the sum to the right of the column. Repeat with each individual column of numbers, working your way from right to left, each time placing the column sum under the previous sum and one place to the left. When you've added all the columns what's left is a simple addition problem, in the example below a column of five four-digit numbers is reduced to adding four two-digit numbers, and by moving each digit one place to the left only single digit addition is then required:

```
    3068              1 9      8 + 4 + 2 + 4 + 1 = 19
    4664            2 1        6 + 6 + 5 + 0 + 4 = 21
    8152          2 1          0 + 6 + 1 + 7 + 7 = 21
    6704      +  2 6           3 + 4 + 8 + 6 + 5 = 26
 +  5741        2 8 3 2 9
   28,329
```

Just as in the previous two examples the column sums you write beside the problem give you a neat and tidy way of keeping track of your work. You could just as quickly sum the individual columns and carry the tens digit over to the next column but this can get messy, and this method has the added benefit of giving you a nice visual way of checking your answer. Let's try one more:

```
    4136                    2 9      6 + 6 + 2 + 3 + 8 + 4 = 29
    6416                  1 2        3 + 1 + 5 + 0 + 3 + 0 = 12
    7952                1 6          1 + 4 + 9 + 1 + 0 + 1 = 16
    3103          +   3 3            4 + 6 + 7 + 3 + 8 + 5 = 33
    8038              3 4 7 4 9
  + 5104
   34,749
```

The longer the column of numbers the more useful this method can be.

A GEOMETRIC RIDDLE

An officer had a legion of men

Who in formation were just 12 by 10,

But without reinforcements his men he could place

In 12 separate rows and evenly spaced,

And 11 in each of the rows would be ranked

Equally distant from the officer they flanked.

How did the officer without men to spare

Seemingly make some appear from thin air?

More Left Hand Addition

This method builds on the previous tip of adding up from the bottom by adding multiple columns. In the example below we start at the bottom left and add 78 to the 9 above to get 87, add 87 to 80 to get 167, etc., and working our way up gives us a total of 281. Starting from the bottom right we add 99 to 7 to get 106, 106 + 60 = 176, etc., working our way up to the top gives us 352. We add 281 to 352 with the 1 in 281 over the 3 in the 352 as shown below:

```
   1798
   9788
   8967
 + 7899
   281
 +  352
  28452
```

Step 1: Split the four columns up into two columns of two.

Step 2: Starting with the left column of two digits add the bottom two numbers to their ones digit above, then the tens digit, and repeat until you get to the top of the column. Put the column sum below the left two-digit column.

Step 3: Repeat Step 2 for the right column of two digits. Put the first digit of this sum under the last digit of the last sum.

Step 4: Add the two numbers to get the sum of all the numbers

This also works with columns of 6 or more:

```
       636251
       420769
       980364
       415627
       198995
  +    118991
       274
          306
  +         397
       2770997
```

Step 1: Split the six columns up into three columns of two.

Step 2: Starting with the left column of two digits add the bottom two numbers to their ones digit above, then the tens digit, and repeat until you get to the top of the column. Put the column sum below the left two-digit column.

Step 3: Repeat Step 2 for the middle column of two digits. Put the first digit of this sum under the last digit of the last sum.

Step 4: Repeat Step 3 for the right column of two digits. Put the first digit of this sum under the last digit of the last sum.

Step 5: Add the three numbers to get the sum of all the numbers.

Easier Subtraction by Rounding

The next time you need to do subtraction try rounding one of the numbers up to the next 10 or 100, add to the other number the same amount, then subtract:

```
  97 + 2 =    99
- 38 + 2 = - 40
             ───
              59
```
Round 38 up to 40 and add 2 to 97 and you have a much simpler subtraction problem.

```
  83 + 5 =    88
- 65 + 5 = - 70
             ───
              18
```
Round 65 up to 70 and add 5 to 83 and now you have a problem you can do in your head!

```
  637 + 12 =    649
- 388 + 12 = - 400
               ───
                249
```
This also works with 3-digit numbers!

```
  732 + 24 =    756
- 476 + 24 = - 500
               ───
                256
```

I know it may not make sense to add to a subtraction problem to make it easier but in the example above it should be fairly easy to see 24 can be added to 476 to raise it to the next hundred. You should also be able to add 24 to 732 in your head, because you are basically adding 2 + 3 and 2 + 4 which gives you the 5 and 6 in 756.

This also works for addition problems. 764 + 386 can be made easier by adding 14 to both numbers to get 778 + 400 which equals 1,178.

This method will take some practice but once you get the hang of it you'll be surprised at how quickly you can do small subtractions.

A CLUE FOR MATH RIDDLE #82

A carpenter had a piece of lumber
Whose sides were of an even number,
The short side measured 2 and the long side 10
But with just 4 cuts he could make them even.
He first cut a square and then some triangles,
Facing inward he placed the right angles.
Then in the middle a square he would place
And outward he made the hypotenuse face.
This is how to make the sides the same size,
Can you a similar styled square devise?

Subtraction by Inspection

Another way to simplify subtraction is to leave out the digits in the ones place and subtract the whole numbers, then subtract the ones digit from the subtrahend (the number on the bottom doing the subtracting) and add the ones digit from the minuend (the number on top being subtracted from). For example, if we wanted to subtract 39 from 73 we would leave out the 3 and 9 and subtract 70 from 30 to get 40, then subtract the 9 to get 31 and add the 3 to get 34:

$73 - 39 = 70 - 30 = 40 - 9 + 3 = 34$
$97 - 38 = 90 - 30 = 60 - 8 + 7 = 59$
$98 - 37 = 90 - 30 = 60 - 7 + 8 = 61$

Subtract the whole numbers, then subtract the ones digit from the bottom, and add the ones digit from the top.

This may take a little practice, but once you get the hang of it you should be able to do all small subtractions like this in your head. Let's try a few harder ones:

$273 - 139 = 270 - 130 = 140 - 9 + 3 = 134$
$397 - 238 = 390 - 230 = 160 - 8 + 7 = 159$
$498 - 137 = 490 - 130 = 360 - 7 + 8 = 361$

*Subtract the whole numbers,
then subtract the ones digit from the bottom,
and add the ones digit from the top.*

You could have also broken up the subtraction problems above into hundreds first and then done the two-digit subtraction, it's whatever you think is easier:

273 – 139 = 200 – 100 = 100, 70 – 30 = 40 – 9 + 3 = 134

397 – 238 = 300 – 200 = 100, 90 – 30 = 60 – 8 + 7 = 159

498 – 137 = 400 – 100 = 300, 90 – 30 = 60 – 7 + 8 = 361

**Subtract the whole numbers,
then subtract the ones digit from the bottom,
and add the ones digit from the top.**

DID YOU KNOW?

The Greek symbol π was first used by the English mathematician William Jones in 1706 but it did not come into general use until about the middle of the 18th century. William Jones chose π because it is the Greek initial letter of the word "periphery". Before 1706 π was simply referred to as "that circle thing".

How to Multiply Any Number by 11

To multiply any number by 11 write down the first digit of the multiplicand (the thing being multiplied) for the first digit of the answer, add the next digits of the multiplicand for the next digits of the answer, then write down the last digit of the multiplicand to complete the answer:

$$25 \times 11 = 275 \quad 2, \text{ then } 2 + 5, \text{ then } 5$$
$$32 \times 11 = 352 \quad 3, \text{ then } 3 + 2, \text{ then } 2$$
$$44 \times 11 = 484 \quad 4, \text{ then } 4 + 4, \text{ then } 4$$

For larger numbers you'll need to carry the 1, this will allow you to multiply numbers of any length by 11 without a calculator.

Take $75,794 \times 11$ for example, all the numbers to be added together will be greater than 9 so we'll have to carry a 1 each time except for the last two digits. This isn't a big deal because we can just add the numbers like we did in the examples above and add 1 to each number.

The first number is 7 which will be 8 when we carry 1.

The next two numbers are $7 + 5 = 12$ and $5 + 7 = 12$ so they will both be 3 after we carry 1.

The fourth number is $7 + 9 = 16 + 1 = 17$, so the fourth number is 7.

The next number is 9 + 4 = 13, so the fifth number is 3 and the last number is 4.

So the answer is a six-digit number, 833,734, and now you can do this kind of problem entirely in your head!

Let's try one more with a mixed bunch of numbers, 325,786 x 11.

The first digit is 3 and we see the next number is a 2 so we won't need to carry a 1.

Second digit is 3 + 2 = 5 and we see the next number is a 5 so again we won't need to carry a 1.

Third digit is 5 + 2 = 7 but the next digit is a 7 which will make the third digit an 8.

Fourth digit is 7 + 5 = 12, and we've already carried the one, but the next number is an 8 so this digit will be a 3.

Fifth digit is 8 + 7 = 15 and we see the next digit is a 6 which will make the fifth digit a 6.

Sixth digit is 6 + 8 = 14 and we carried our 1 already, and since it is the second to last number we never worry about having to add a 1.

Seventh digit is a 6, no extra work necessary.

So we have 3,583,646 as our answer, no calculator needed!

By itself this tip may not seem too terribly useful but as you learn more tips you'll see how combining tips can help you quickly and easily solve a wide variety of math problems.

DID YOU KNOW?

To compute π means to calculate how many times larger the circumference of a circle is to its diameter. This calculation is called "the numerical rectification of a circle". The Hindu mathematician Aryabhata (476 - 550 AD) calculated π even greater than Archimedes and found the ratio 62,832/20,000 or 3.1416. Hindu mathematicians continued to double the sides of the polygon until they reached 384 and found the ratio 3,927:1,250. Pretty impressive when you consider the fact they were doing this all manually with ancient protractors and whatever they used for pencils and paper back then.

Quickly Multiply Numbers Ending in 5

When your multiplier (the number on the bottom) and multiplicand (number on the top) both end in 5 there is a simple trick easy enough for you to do in your head. The last two numbers will always be either 25 or 75, and the way to tell which number to use is to add all the numbers together except for the 5's, and if the sum is even the answer will end in 25 and if the sum is odd you use 75. Then multiply all but the 5's and add half their sum, rounding down if your answer will end in 75. Let's look at an example:

```
    225
  x  65
  14,625
```

Step 1: 22 + 6 = 28 which is even so the answer ends in 25

Step 2: 22 x 6 = 132, add ½ of 28: 132 + 14 = 146

Now let's do it the normal way:

```
    225
  x  65
  1,125
 13,500
 14,625
```

This way requires 6 multiplications and 10 additions!

Let's try a few more:

18 x 9 = 162
162 + (27/2 = 13) = 175

185
x 95
17,575

18 + 9 = 27
which is odd
so the
answer ends
in 75

8 x 7 = 56
56 + (15/2 =7) = 63
so our 1st two
numbers are 63

85
x 75
6,375

8 +7 = 15
which is
odd so it
ends in 75

6 x 4 = 24
24 + (10/2 =5) = 29
So it starts with
29

65
x 45
2,925

6 + 4 = 10
which is
even so it
ends in
25

11 x 7 =77
77 + (18/2 =9) = 86
So it starts with
86

115
x 75
8,625

7 +11 = 18
which is
even so it
ends in
25

Quickly Square Numbers Ending in 5

This tip is even easier than the last. To square any number ending in 5 multiply the number on the left by the next higher number, for example 65^2 would be 6 x 7 = 42 then add a 25 to get the answer 4,225. Let's compare our new tip to the normal way of doing this kind of problem:

```
    65
  x 65
   325      This way requires 4
  3900      multiplications and
  4,225     7 additions!
```

Let's try a few more:

15^2 = 1 x 2 = 2 and 25 = 225

25^2 = 2 x 3 = 6 and 25 = 625

35^2 = 3 x 4 = 12 and 25 = 1,225

45^2 = 4 x 5 = 20 and 25 = 2,025

55^2 = 5 x 6 = 30 and 25 = 3,025

Three-digit numbers are a little harder, it helps to mix rules like multiplying by 11:

105^2 = 10 x 11 = 110 and 25 = 11,025

115^2 = 11 x 12 = 132 and 25 = 13,225

If we wanted to square 205 we would multiply 20 x 21, or we could just multiply 20 x 20 and add the other 20 back in:

205^2 = 20 x 21 or 20 x 20 + 20 = 420 and 25 = 42,025

405^2 = 40 x 41 or 40 x 40 + 40 = 1,640 and 25 = 164,025

DID YOU KNOW?

Some more π facts: In 1585 the Dutch geometer and astronomer Adriaan Metius placed the number between 377/170 and 333/106 and found his famous fraction 355/113. No two numbers more exactly represent the value of π than the ratio 355:113 which is 3.14159292, within a third of a millionth of the actual number, 3.14159265.

A Quick Way to Multiply When Your Multiplier Ends in 1

When your multiplier ends in 1 drop the 1 and multiply, then add the multiplicand to your answer you've shifted one place to the left. Let's say we want to multiply 23,425 by 41:

Drop the 1 and multiply by 4

Shift the multiplicand one place to the right and add the previous product

$$\begin{array}{r} 23{,}425 \\ \times \quad 4 \\ \hline 93{,}700 \\ \\ 23425 \\ +93700 \\ \hline 960425 \end{array}$$ **2 steps!**

The normal way:

$$\begin{array}{r} 23{,}425 \\ \times \quad 41 \\ \hline 23425 \\ 937000 \\ \hline 960{,}425 \end{array}$$ **3 steps!**

MATH FUNNIES

$\sqrt{-1}$ 2^3 Σ π

and it was delicious!

A Quick Way to Multiply When Your Multiplier Ends in 9

When your multiplier ends in 9 round up to the nearest 10 and subtract the multiplicand. For example, if we wanted to multiply 327 by 39 we would round 39 up to 40 and then subtract 327:

Round up to 40

Subtract the multiplicand

$$\begin{array}{r} 327 \\ \times\ \ 40 \\ \hline 13,080 \\ -\ 327 \\ \hline 12,753 \end{array}$$

2 steps!

The normal way:

$$\begin{array}{r} 327 \\ \times\ \ 39 \\ \hline 2943 \\ 9810 \\ \hline 12,753 \end{array}$$

This way requires 5 multiplications and 6 additions!

This also works the other way, if our multiplier ends in 1 we could round down to 40 and <u>add</u> the multiplicand:

Round down to 40

add the multiplicand

$$\begin{array}{r} 327 \\ \times\ \ 40 \\ \hline 13,080 \\ +\ 327 \\ \hline 13,407 \end{array}$$

2 steps!

A Quick Way to Multiply When the Multiplier is a Composite Number

When the multiplier is a composite number like 42 multiply by its factors 6 and 7 instead. This may seem like more work but it eliminates one step of the multiplication:

$$\begin{array}{r} 328 \\ \times\ \ 6 \\ \hline 1,968 \\ \times\ \ 7 \\ \hline 13,776 \end{array}$$ *2 multiplications, No addition necessary*

And the normal way:

$$\begin{array}{r} 328 \\ \times\ 42 \\ \hline 656 \\ 13120 \\ \hline 13,776 \end{array}$$ *This way requires 6 multiplications and 6 additions!*

This also works for larger numbers, how about we try this by multiplying 4,786 by 112. We divide 112 by 2 as the first step to finding its factors, and 112 ÷ 2 = 56, and we know the factors of 56 are 7 and 8, so we'll multiply 4,786 by 2, 7, and 8:

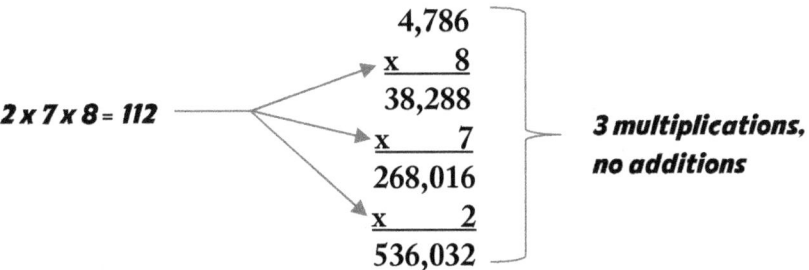

The normal way:

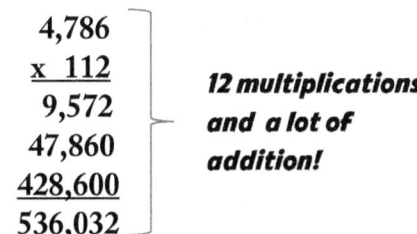

DID YOU KNOW?

Here is a way to remember π to 12 places:
"See I have a rhyme assisting my feeble brain its tasks sometimes resisting". The number of letters in each word gives the digits for π. You should be able to impress everyone with your π knowledge on π Day, 3/14!

A Quick Way to Multiply by 15, 150, & 1,500

When your multiplier is 15, 150, or 1,500 add a zero to the multiplicand and add half the number:

$$324 \times 15 = \begin{array}{r} 3{,}240 \\ +\,1{,}620 \\ \hline 4{,}860 \end{array}$$

**Half of 324 is 162
add a zero to both**

$$324 \times 150 = \begin{array}{r} 32{,}400 \\ +\,16{,}200 \\ \hline 48{,}600 \end{array}$$

**add 2 zeros for 150
only 2 steps!**

$$324 \times 1{,}500 = \begin{array}{r} 324{,}000 \\ +\,162{,}000 \\ \hline 486{,}000 \end{array}$$

add 3 zeros for 1,500

The normal way:

$$\begin{array}{r} 1{,}500 \\ \times\ 324 \\ \hline 6{,}000 \\ 30{,}000 \\ 450{,}000 \\ \hline 486{,}000 \end{array}$$

12 multiplications and a lot of addition!

You might be thinking to yourself, this is all well and good but it's kind of hard to divide some numbers by 2. Well, there's a tip for that too. If you break a number up into 2 or 3 parts it's much easier to find one half:

$$3{,}240 = 3{,}000 + 240$$
$$3{,}000 \div 2 = 1{,}500$$
$$240 \div 2 = 120$$
$$1{,}500 + 120 = 1{,}620$$

We could have also taken this one step further and said 3,240 was equal to 3,000 + 200 + 40, it's whatever is easier for you.

Some of you may be thinking "What about the odd numbers? How do we find half of those?" The answer is there will never be any odd numbers because when we add zeros we make them all even numbers.

DID YOU NOW?

Leonardo Fibonacci convinced the Western world Roman numerals were best used only for fancy clock faces in 1202 in his bestselling book *Liber Abaci*. After studying with Arab mathematicians he brought back to Italy the idea of *modus Indorum* (method of the Indians), the Hindu-Arabic numerals we use today. Before then mathematicians struggled to do math with Roman numerals. What's CXXXVII + CCLXXIII? That would be 137 + 273. Thanks, Leonardo!

How to Multiply by Complements

When your multiplicand and multiplier are relatively close to the next value of a hundred find the numbers' "complement", or how far they are away from the next hundred value, and use it to help you find the product. To multiply 98 by 95 you would write down a 2 and 5 to bring 98 and 95 to 100, multiply those numbers to get the right digits of your answer, and subtract either of the complements from the opposite number to get the left digits. It doesn't matter which one you use as long as you subtract the opposite number because you get the same answer either way.

$$\begin{array}{r} 98 \\ \times\ 95 \\ \end{array} \quad \begin{array}{l} 100 - 98 =\ \ 2 \\ 100 - 95 = \times\ 5 \\ \hline 10 \end{array}$$

$$98 - 5 = 93$$
$$95 - 2 = 93$$

The answer will be four digits since you multiplied two digit numbers so now all you have to do is put 93 and 10 together to get the answer, 9,310.

Step 1: Multiply your complements together to get the right 2 digits

Step 2: Subtract one of the numbers from the other complement to get the left digits, it doesn't matter which one, the answer is the same either way!

This also works with bigger numbers, you just need to correct the left digits by multiplying them by the value of the hundred you're complementing.

- Between 100 and 200 double the left number
- Between 200 and 300 triple the left number
- Between 300 and 400 quadruple the left number
- Etc.

This will work for numbers from 50 to 99 for any value of a hundred as long as both numbers have the same hundred value and both are over 50. For example you could use this method to multiply 351 by 362:

$$\begin{array}{ll} 351 & 400 - 351 = 49 \\ \underline{\times 362} & 400 - 362 = \underline{\times\ 38} \\ & 1{,}862 \end{array}$$

$$351 - 38 = 313$$
$$362 - 49 = 313$$

This is where the other tips come in handy. You can use the method of multiplying any two numbers to multiply 49 by 38 which with a little practice you can do in your head. The subtraction and addition tips can be used to simplify the process as well.

The hundred you are complementing is 400 so multiply 313 by 4 to get the left digits:

$$313 \times 4 = 1{,}252$$

If you multiply 400 x 400 you get 160,000 so you know your answer will be 6 digits and less than 160,000. Since the answer is 6 digits and you have two 4 digit numbers slide the right numbers over to get a 6 digit answer:

```
   1,252
+  1,862
 127,062
```

This may seem like a lot of work but with smaller numbers you should be able to solve for the answer entirely in your head. More complicated problems like the previous example still require fewer steps than doing it the normal way which requires 9 multiplications and a lot of addition. The complements method only requires 1 or 2 multiplications (which you should be able to do in your head) and a little bit of subtraction and 1 addition.

Let's work through a few more so you can get the hang of it:

```
   198   200 - 198 =    2
 x 195   200 - 195 =  x 5
 38,610                10
```
195 - 2 = 193 x 2 = 386
The left three numbers are 386
The right two numbers are 10

```
   293   300 - 293 =    7
 x 287   300 - 287 =  x 13
 84,091                91
```
287 − 7 = 280 x 3 = 840
The left three numbers are 840
The right two numbers are 91

```
    496   500 - 496 =    4        483 - 4 = 479 x 5 = 2,395
x   483   500 - 483 = x 17        The left 4 numbers are 2,395
239,568                 68        The right two numbers are 68


    594   600 - 594 =    6        578 - 6 = 572 x 6 = 3,432
x   578   678 - 578 = x 22        Add 3,432, to 132:
343,332                132
                                    3,432
                                  +  132
                                  343,332


     87   100 - 87 =    13        87 - 11 = 76
 x   89   100 - 89 =  x 11        Add 76 to 143:
 7,743                 143
                                     76
                                  +  143
                                  7,743


     51   100 - 51 =    49        51 - 38 = 13
 x   62   100 - 62 =  x 38        Add 13 to 1,862:
 3,162                1,862
                                     13
                                  + 1,862
                                   3,162


  1,072  1,100 - 1,072 =  28      1,072 - 14 = 1,058 x 11 = 11,638
x 1,086  1,100 - 1,086 = x14      Add 11,638 to 392:
1,164,192                392
                                   11,638
                                  +   392
                                  1,164,192
```

How to Multiply by Excesses

A variation of multiplying by complements is multiplying by excesses. If your numbers are the same hundred value and less than 50 you can use their "excesses", the amount they are over the hundred value, to find the product. Subtract the excess of the multiplier from the multiplicand, multiply both excesses, and put the numbers together to get the answer.

Let's multiply 112 by 103 using this method:

```
  112     Step 1: Take the excess of the      112
x 103     multiplier and add it to the        + 3
          multiplicand:                       115

          Step 2: Take the excesses from       12
          the multiplier and the              x 3
          multiplicand and find the            36
          product of the two:

          Step 3: Put 115 and 36 together    11,536
          for the answer.
```

The normal way:

6 multiplications and a lot of addition!

The multiplication by excesses method only took three steps, or two if you don't count putting the two numbers together. This also works with bigger numbers but this time you correct the left digits by multiplying them by the value of the hundred you're excessed. If you're number is over 200 multiply by 2, over 300 multiply by 3, etc. If your number is over 1,000 you would multiply by 10 so just add a zero to the left numbers.

Let's try a few more so you can get the hang of it:

1,009	1,009	1,016	9	10,160
x 1,007	+ 7	x 10	x 7	+ 63
	1,016	10,160	63	1,016,063

132	132		32	140
x 108	+ 8		x 8	+ 256
	140		256	14,256

324	324	340	24	1,020
x 316	+ 16	x 3	x 16	+ 384
	340	1,020	384	102,384

1,013	1,013	1,025	13	10,250
x 1,012	+ 12	x 10	x 12	+ 156
	1,025	10,250	156	1,025,156

507	507	519	12	2,595
x 512	+ 12	x 5	x 7	+ 84
	519	2,595	84	259,584

Quickly Multiply Alternating Numbers

Alternating numbers have a common number between them, like 15 and 17, 24 and 26, and 39 and 41. To quickly multiply them square the common middle number, or the "intermediate" number, and then subtract 1:

$$15 \times 17 = 16^2 - 1 = 256 - 1 = 255$$
$$17 \times 19 = 18^2 - 1 = 324 - 1 = 323$$
$$39 \times 41 = 40^2 - 1 = 1{,}600 - 1 = 1{,}599$$

This also works for numbers having three intermediate numbers, you just need to subtract 4 in this case:

$$18 \times 22 = 20^2 - 4 = 400 - 4 = 396$$
$$23 \times 27 = 25^2 - 4 = 625 - 4 = 621$$
$$48 \times 52 = 50^2 - 4 = 2{,}500 - 4 = 2{,}496$$

DID YOU KNOW?

Our number system is base-10 (decimal) most likely because we have 10 fingers and toes. This means in the Simpsons' world and on ET's planet their number systems are probably base-8 (octal) because they only have 8 fingers!

Quickly Square Any Number of Two Digits

To square any two-digit number square the number on the right, take twice the product of the two numbers for the second digit, then square the number on the left for the left digits, remembering to add in the numbers you carried over each time.

Let's try it with 86^2:

Step 1: Square the ones digit and use the ones digit of the answer for the rightmost digit and carry the rest.

$6^2 = 36$, right number is 6, carry 3.

Step 2: Take two times the product of both numbers and use the ones digit of the answer for the next number, carry the rest.

$2(8 \times 6) = 96$, next number is $6 + 3 = 9$, carry 9

Step 3: Square the first digit and add the carry from the previous step for the solution.

$8^2 = 64 + 9 = 73$

The answer is 7,396

Let's try 74^2:

Step 1: $4^2 = 16$, rightmost digit is 6, carry 1

Step 2: $2(7 \times 4) = 56$, second digit is $6 + 1 = 7$, carry 5

Step 3: $7^2 = 49 + 5 = 54$

Put 54 and 7 and 6 together to get the answer 5,476.

How about 83^2:

Step 1: $3^2 = 9$, rightmost digit is 9

Step 2: $2(8 \times 3) = 48$, second digit is 8, carry 4

Step 3: $8^2 = 64 + 4 = 68$

We put together 68 and 8 and 9 and the answer is 6,889.

By breaking the problem up into three smaller problems we've simplified the process to the point where you should be able to square most two-digit numbers in your head.

Now let's try these the normal way:

4 multiplications each and a lot of additions!

```
37² = 37         83² = 83         74² = 74
      x 37             x 83             x 74
       259              249              296
     1,110            6,640            5,180
     1,369            6,889            5,476
```

This also works with three-digit numbers but it's a little more complicated – let's look at 313^2:

Step 1: $3^2 = 9$, right number is 9

Step 2: $2(31 \times 3) = 2(93) = 186$, 2nd number is 6 carry 18

Step 3: $1^2 = 1 + 8$ from the carry, 3rd number is 9 carry 10

Step 4: $2(3 \times 1) = 2(93) = 6 + 1$ from the carry, 4th number is 7

Step 5: $3^2 = 9$, 5ᵗʰ number is 9

So the answer is 97,969. The trick is to treat it like two separate problems using the same method, the first time you multiply diagonally the numbers are 6 x 31 and the second time the numbers are 1 x 3. Work your way from right to left and with a little practice you will be able to do problems like this in your head. It also helps to combine tips like multiplying by 11, numbers ending in 5 or 15, etc.

DID YOU KNOW?

Most people associate the discovery of calculus to Sir Isaac Newton but German mathematician Gottfried Liebniz published a paper explaining his invention of calculus in 1684, three years before Newton mentioned it in his *Philosophiæ Naturalis Principia Mathematica* (which is usually just referred to as the *Principia*). Liebniz died in 1716 before the feud with Newton could be settled. This was probably the most drama the math world had ever experienced. The integral symbol, ∫, is the long S symbol Liebniz used in his calculations which he chose because he thought of the integral as "the infinite sum of infinitesimal summands" (a fancy way to say addition problems).

How to Find the Sum of Any Sequence of Numbers

To calculate the sum of a list of numbers in your head you first need to do something that doesn't make sense, double the list and reverse the numbers. Let's try this with the numbers from 1 to 10:

	1	2	3	4	5	6	7	8	9	10
+	10	9	8	7	6	5	4	3	2	1
	11	11	11	11	11	11	11	11	11	11

When we double the list but reverse the second list each of the sums of the consecutive numbers always equals 11. To find the sum of all the numbers we just multiply 11 by 10 to get 110 and divide by 2 since we had to double the list which gives us 55.

Let's see if we're right:

$$1 + 2 + 3 + 4 + 5 + 6 + 7 + 8 + 9 + 10 = 55$$

This works for any list of numbers if we use the formula

$$\frac{n \times (n+1)}{2}$$

Now we can work with a bigger list like 1 to 100:

$$\frac{100(100+1)}{2} = \frac{100 \times 101}{2} = \frac{10,100}{2} = 5,050$$

When you sum series of multiples of 10 the answer always adds up to the number. For example summing the numbers from 1 to 10 is 55, and 5 + 5 = 10. Summing the numbers from 1 to 100 is 5,050, and 50 + 50 = 100.

What do you think the sum of the numbers from 1 to 1,000 might be? My bet is 500,500, how about you? We learned from the formula we first need to multiply 1,000 by 1,001, and a quick way to do this is to drop the 1 and then to multiply 1,000 by 1,000. An easy way to do this is to write down 1 and add all the zeros to it – 1,000 and 1,000 have 6 zeros between the two so the answer is 1,000,000. Adding in the 1,000 you left out gives you 1,001,000, so 1,000 x 1,001 = 1,001,000. To divide by 2 we can break this into two problems, 1,000,000 ÷ 2 = 500,000 and 1,000 ÷ 2 = 500, so the answer to our formula is 500,500, and 500 + 500 = 1,000!

We represent this mathematically as:

$$\sum_{i=1}^{n} i$$ *We call this summation notation!*

where *n* is the number we're summing to and *i* is the starting number, and if we wanted to represent summing the numbers from 1 to 100 we would express it as:

$$\sum_{i=1}^{100} i$$

where *n* is 100 in this case and we're starting with the number 1. The symbol Σ is the upper-case Greek letter Sigma, the eighteenth letter of the Greek alphabet, and carries the 's' sound. Σ is used in mathematics to represent summation.

This will work for any sequence of numbers, let's try to sum the numbers from 1 to 18:

$$\sum_{i=1}^{18} i$$

$$\frac{18(18+1)}{2} = \frac{18 \times 19}{2} = \frac{342}{2} = 171$$

A quick check with the calculator tells us the answer is indeed 171. We could have used two of our previous tips to help multiply 18 x 19, one way would have been to round 19 to 20:

```
     18                              360
   x 20         Then subtract 18:   - 18
    360                              342
```

Or, we could have rounded 18 up to 19 and found 19^2:

Step 1: $9^2 = 81$, right number is 1 carry 8

Step 2: 2(1 x 9) = 18, 8 + 8 = 26, 2nd number is 6 carry 2

Step 3: 1^2 = 1 + 2 = 3

So 19^2 = 361, and 361 – 19 = 342.

What? Subtracting 19 from 361 is too hard? Remember our subtraction tip:

361 + 1 =	362	**Round 19 up to 20 and add 1 to 361 and you have a much simpler subtraction problem.**
– 19 + 1 =	– 20	
	342	

To divide 342 by 2 we can break it up into 300 ÷ 2 which is 150, and 42 ÷ 2 which is 21, and 150 + 21 = 171.

Another way to do this is to multiply half the sum of the first and last numbers of the series by the number of terms, so for 1 to 18 the answer is:

$$18\left(\frac{1+18}{2}\right) = 18\left(\frac{19}{2}\right) = 18 \times 9\frac{1}{2} = 171$$

To multiply 18 x 9 ½ in your head try this:

18 is 9 + 9, so if we break this into two separate problems and ignore the ½ for a moment 18 x 9 is the same as:

$$(9 \times 9) + (9 \times 9)$$

you multiply 9 x 9 twice and add them together to get:

$$81 + 81 = 162$$

(or 80 + 80 + 1 + 1, another shortcut from a previous tip). Then factor in the ½ by taking ½ of 18 which is 9, and add it to 162 to get 171.

There are many, many ways we can simplify mathematics if we use a little imagination. Now let's learn how to sum geometric series.

DID YOU KNOW?

A 20,000 year old artifact discovered in Africa called the Ishango bone shows early man had a concept of time and numbers. The bone has a series of marks carved in three columns running the length of the bone and is thought to be the earliest known demonstration of sequences of prime numbers or a six-month lunar calendar.

How to Calculate the Sum of a Geometric Series

A geometric series is one where each number is found by multiplying its previous number by a common number, also called the *common ratio*, for example:

1, 2, 4, 8, 16, 32, 64, 128, 256 … common ratio is 2.

1, 3, 9, 27, 81, 243, 729, 2,187 … common ratio is 3.

To find the sum of a geometric series multiply the last number by the common ratio, subtract the lowest number of the series from the product, and divide by the common ratio less one. Let's try to find the sum of our first series with a common ratio of 2 using the numbers from 1 to 16 to test our formula:

$$\frac{2(16)-1}{2-1} = \frac{32-1}{1} = 31$$

See, this sounded a lot more difficult than it really is. Let's use the second example geometric series with a common ratio of 3 and find the sum of the numbers from 1 to 81:

$$\frac{3(81)-1}{3-1} = \frac{243-1}{3-1} = \frac{242}{2} = 121$$

This is even easier for a series of numbers with an odd number of numbers – just count the numbers in the series and multiply that number by the middle number. For example, 1 + 2 + 3 + 5 + 7 has

5 numbers and the middle number is 3 so the sum of the series is 15.

1 + 3 + 5 + 7 + 9 also has 5 numbers but its middle number is 5 so the sum of the series is 5 x 5 = 25. Why does this work? The middle number is always the average of the numbers, and the average is found by adding the numbers and dividing by the number of terms. Here we are just working backwards to find the sum of the series.

DID YOU KNOW?

The earliest evidence of written mathematics dates back to the ancient Sumerians who built the earliest civilization in Mesopotamia. They developed a complex system of measurement around 3000 BC and about 2500 BC they wrote multiplication tables, geometry, and division problems on clay tablets. Some of these tablets appear to be graded school work which means Sumerian math students couldn't say their dog ate their homework!

Division Tips

1. To divide by 2½ multiply the dividend by 4 and move the decimal point one place to the left. To divide 100 ÷ 2½ simply multiply 10.0 x 4 to get the answer 40. Or you could do it the long way:

$$\frac{100}{2\frac{1}{2}} = \frac{100}{\frac{5}{2}} = \frac{100 \times 2}{5} = \frac{200}{5} = 40$$

Dividing by a fraction is the same as multiplying by the reciprocal of the same fraction, this is why we multiplied 100 x 2 and divided by 5.

Why does this work? The decimal of ¼ is .25 and moving the decimal one place to the right makes it 2.5. When a divisor is a fraction you flip it, or take its reciprocal, and multiply it by the dividend (the number on top). If we substitute 2.5 with ¼ and move the decimal of the dividend one place then you will get the same result as dividing by 2½. Once we do a few more this will make perfect sense.

2. To divide by 25 multiply the dividend by 4 and move the decimal point *two* places to the left. For example, if we wanted to divide 375 ÷ 25 you would multiply 3.75 x 4 to get 15. Again, this is because the decimal of the fraction ¼ is .25 which is the same as dividing by 25 and moving the decimal two places to the left.

3. To divide by 250 multiply the dividend by 4 and move the decimal point *three* places to the left. Now you can see a trend, dividing by 2.5 move the decimal one place, by 25 move the decimal two places, by 250 three places, and so on. You move the decimal the same number of places as there are digits in the number.

4. To divide by 5 multiply the dividend by 2 and move the decimal point one place to the left. To divide 100 ÷ 5 the answer would be 10.0 x 2 = 20. Let's try a harder one: for 360 ÷ 5 just multiply 36 x 2 to get the answer 72. Or for 455 ÷ 5 multiply 45.5 x 2 to get 91.

This works because the decimal of $\frac{1}{5}$ is .20 and all you have to do is multiply by 2 and move the decimal one place to get the same result as dividing by 5. Remember, dividing by a fraction is the same as multiplying the dividend by the denominator of the fraction.

5. To divide by 50 multiply the dividend by 2 and move the decimal point *two* places to the left. So 3,750 ÷ 50 is the same as 37.50 x 2 which is 75.

6. To divide by 500 multiply the dividend by 2 and move the decimal point *three* places to the left. I hope you can see a theme

developing here, the number of places you move the digit is equal to the number of digits. 5 was one place, 50 was two places, 500 was three places, and so on.

7. To divide by 1 and any number of zeros move the decimal point the same number of places to the left as there are zeros:

$$\frac{2,384}{10} = 238.4 \qquad \frac{2,384}{100} = 23.84 \qquad \frac{2,384}{1,000} = 2.384$$

8. To divide by 1¼ multiply the dividend by 8 and move the decimal point one place. So, if we wanted to find 100 ÷ 1¼ you just multiply 10.0 x 8 to get 80. Or you could do it the long way:

$$\frac{100}{1¼} = \frac{100}{5/4} = \frac{100 \times 4}{5} = \frac{400}{5} = 80$$

Why does this work? The decimal of ⅛ is .125 so dividing by 1.25 is the same as moving the decimal point one place and multiplying by 8. When the divisor is a fraction multiply the dividend by the denominator of the fraction which in this case is 8 and move the decimal one place. The easy way to remember this is to move the decimal the same number of places as there are positive digits.

9. To divide by 12½ multiply the dividend by 8 and move the decimal point *two* places. If we wanted to find 100 ÷ 12½ you

just multiply 1.00 x 8 to get 8. Let's try a harder one, for 350 ÷ 12½ multiply 3.50 x 8 to get the answer 28. Or how about 525 ÷ 12½, multiply 5.25 x 8 to get the answer 42 (the answer to Life, the Universe, and Everything). The multiplication is easy, 8 x 5 = 40 and 8 x ¼ = 2, and 40 + 2 = 42.

10. To divide by 125 you multiply the dividend by 8 and move the decimal point *three* places to the left. To divide 3,500 ÷ 125 multiply 3.500 by 8 to get 28. This is exactly the same as dividing by 12½, you just need to move the decimal one more place.

11. To divide by 16⅔ multiply the dividend by 6 and move the decimal point two places to the left. If we wanted to find the answer to 100 ÷ 16⅔ we would just multiply 1.00 x 6 to get the answer 6. Or you could do it the hard way:

$$\frac{100}{16⅔} = \frac{100}{⁵⁰⁄_3} = \frac{100 \times 3}{50} = \frac{300}{50} = 6$$

This works because the decimal of ⅙ is .16667, move the decimal two places to the right to get 16 and the decimal of the fraction ⅔ is .667, put them together and you get 16⅔.

12. To divide by 20 multiply the dividend by 5 and move the decimal point two places to the left. If you wanted to find 360 ÷ 20 just multiply 360 x 5 to get 1,800 and move the decimal two

places to get the answer 18. This works because the decimal of ½₀ is .05 and moving the decimal two places to the right is 5.

13. To divide by 200 multiply the dividend by 5 and move the decimal point *three* places to the left

14. To divide by 33⅓ multiply the dividend by 3 and move the decimal point two places to the left. For example 9,876 ÷ 33⅓ is the same as 98.76 x 3 which is 296.28. Why? Because the decimal of ⅓ is .3333 and moving the decimal two places to the right gives us 33.33 or 33⅓.

15. To divide by 66⅔ multiply the dividend by 3, move the decimal point two places to the left, and divide by 2. For example, if you wanted to find 100 ÷ 66⅔ multiply 1.00 x 3 to get 3 and divide by 2 to get 1½. This is the same as dividing by 33⅓ but since 66⅔ is twice as much you have to divide by 2 after you multiply by 3.

Since the decimal of ¹⁄₁₅ is .066667 you could also divide by 66⅔ by moving the decimal three places and multiplying by 15. For example, 3,456 ÷ 66⅔ can be solved by multiplying 3.456 x 15 to get 51.84. To simplify multiplying by 15 use its composites 3 and 5 instead, 3.456 x 3 = 10.368 x 5 = 51.84.

Fraction Tips

1. To add two fractions which have 1 for their numerator write the sum of the given denominators over the product of the given denominators. For example, $\frac{1}{4} + \frac{1}{5} = \frac{9}{20}$. Or, $4 + 5 = 9$ for the numerator and $4 \times 5 = 20$ for the denominator.

2. To subtract two fractions which have 1 for their numerator write the difference of the given denominators over the product of the given denominators: $\frac{1}{4} - \frac{1}{6} = \frac{2}{24} = \frac{1}{12}$.

3. To multiply two mixed numbers when the whole numbers are the same and the sum of the fractions is 1 multiply the whole number by the next highest whole number and add to the product the product of the fractions: $9\frac{4}{5} \times 9\frac{1}{5} = 90\frac{4}{25}$.

4. To multiply two mixed numbers when the difference of the whole numbers is 1 and the sum of the fractions is 1 add 1 to the larger number and multiply it by the smaller number, then square the fraction belonging to the larger number and subtract 1 from its square. Add the whole number and the fraction and you have the desired product.

So for $5\frac{4}{5} \times 4\frac{1}{5}$ add 1 to the first whole number and multiply by the second number to get $(5 + 1) \times 4 = 24$.

Square the larger fraction and subtract 1:

$$(4/5)^2 = 16/25 - 25/25 = 9/25$$

Put them together and the answer is $5\tfrac{4}{5} \times 4\tfrac{1}{5} = 24\tfrac{9}{25}$.

The normal way is $29/5 \times 21/5 = 609/25 = 24\tfrac{9}{25}$ which is a little more work. We can use some of the multiplication tips to make the normal way easier, we could have added 1 to 29 and multiplied by 30 or subtracted 1 from 21 and multiplied by 20, remembering to subtract 21 in the first case or adding 29 in the second case:

```
    29              21   630           29    580
  x 21            x 30  - 21         x 20   + 29
    29             630   609          580    609
   580
   609
```

We should be able to do these in our heads now!

Divide 609 by 25 and since we know $25^2 = 625$ by using another multiplication tip we see the answer to the division problem, or the quotient, has to be 24.

To find 25 x 24 just subtract 25 from 625 to get 600, and you have 9 left over.

5. To multiply two mixed numbers ending in ½ add half the sum of the numbers to their product. If the sum is an odd number, call it one less, to make it even, add ¾.

To multiply 8½ x 6½ you would first multiply 8 x 6 to get 48. Add half the sum of the whole numbers, 8 + 6 = 14/2 = 7, and 48 + 7 = 55, so the whole number of the answer is 55.

Then square the fraction, (½)² = ¼, and your answer is 55¼.

The normal way: 8½ x 6½ = ¹⁷⁄₂ x ¹³⁄₂ = ²²¹⁄₄ = 55¼.

Let's try 5½ x 6½. Multiply 5 x 6 to get 30 and add one half of the sum 5 + 6 but since the answer is odd the fraction will be ¾. Then one half the sum of 5 + 6 is 5 and the answer is 35¾.

6. To square any number ending in one half multiply the number by itself increased by one, and add ¼. So (8½)² is 8 x 9 = 72 plus ¼ which gives us the answer 72¼.

7. To square any number ending in ¼ multiply the number by itself increased by ½ and add ¹⁄₁₆.

If you wanted to find (8¼)² multiply 8 x 8½, 8² = 64, adding half of 8 gives you 68, then it's just a matter of tacking on ¹⁄₁₆ for the answer 68¹⁄₁₆.

The normal way: 8¼ x 8¼ = ³³⁄₄ x ³³⁄₄ = ¹⁰⁸⁹⁄₁₆ = 68¹⁄₁₆.

The normal way is a little more complicated but you can use some of the other tips to make it easier. To multiply 33 x 33 use the tip to square any number of two digits – square the number on the right, multiply the product of the two numbers by 2, then square the number on the left:

Step 1: $3^2 = 9$, rightmost digit is 9

Step 2: $2(3 \times 3) = 18$, second digit is 8, carry 1

Step 3: $3^2 = 9 + 1 = 10$

Put 10, 8, and 9 together to get the answer 1,089.

8. To square any number ending in ¾ multiply the number by itself increased by 1½ and add ⁹⁄₁₆. To find $(8¾)^2$ multiply 8 x 9½, and 8 x 9 = 72 and adding half of 8 gives you 76. Then it's just a matter of tacking on ⁹⁄₁₆ to get the answer 76⁹⁄₁₆.

The normal way: 8¾ x 8¾ = ³⁵⁄₄ x ³⁵⁄₄ = ¹²²⁵⁄₁₆ = 76⁹⁄₁₆

You can simplify the normal way with another multiplication tip. To multiply 35 x 35 use the tip to square numbers ending in 5. When the multiplicand and the multiplier both end in 5 the product always ends in 25.

Then multiply the left digit by 1 number more, 3 x 4 = 12 and adding 25 gives you 1,225. Then it's just a matter of dividing 1,225 by 16.

9. To square any number ending in ⅓ multiply the number by itself increased by ⅔ and add ⅑.

To find $(8\frac{1}{3})^2$ multiply $8 \times 8\frac{2}{3}$. 8 is the same as $\frac{24}{3}$ and $8\frac{2}{3}$ is $\frac{26}{3}$ and $\frac{24}{3} \times \frac{26}{3} = \frac{624}{9}$ plus $\frac{1}{9}$ which gives you the answer $69\frac{4}{9}$.

You can use the multiplication of alternating numbers tip to multiply 24 by 26, 24 x 26 is the same as $25^2 - 1 = 625 - 1 = 624$. Th en it's just a matter of adding $\frac{1}{9}$.

You can use the multiplication of numbers ending in 5 tip to make the normal way easier too. $8\frac{1}{3}$ is $\frac{25}{3}$ and to find 25 x 25 multiply 2 x 3 to get 6 for the first number and tack 25 on the end:

$$8\frac{1}{3} \times 8\frac{1}{3} = \frac{25}{3} \times \frac{25}{3} = \frac{625}{9} = 69\frac{4}{9}$$

10. To square any number ending in $\frac{2}{3}$ multiply the whole number increased by 1 by the whole number plus $\frac{1}{3}$ and add $\frac{1}{9}$.

To square $8\frac{2}{3}$ then you would multiply $8 + 1 \times (8 + \frac{1}{3})$:

$$8 + 1 \times 8\frac{1}{3} = \frac{27}{3} \times \frac{25}{3} = \frac{675}{9} = 75 + \frac{1}{9} = 75\frac{1}{9}$$

The benefit of doing it this way is you always end up multiplying alternating numbers, in this case 24 x 28 is the same as $26^2 - 1$ and you can use the tip for squaring any two digit number to find the answer.

11. To multiply two numbers ending with the same fraction to the product of the whole numbers add that fraction of their sum and the square of the fraction.

This is a three step process, for $15\frac{2}{7} \times 6\frac{2}{7}$ the first step is to find the product of the whole numbers which is $15 \times 6 = 90$.

The sum of the whole numbers multiplied by the fraction is $\frac{2}{7}(15+6)$ which is equal to $\frac{2}{7}(21) = 2 \times 21 = 42 \div 7 = 6$. $90 + 6 = 96$ so our whole number is 96.

Squaring the fraction $\frac{2}{7}$ gives us $\frac{4}{49}$ so our answer is $96\frac{4}{49}$.

12. To square any mixed number multiply the whole number by itself increased by twice the fraction and add the square of the fraction.

For example, for $(8\frac{1}{7})^2$ we would multiply 8 by twice the fraction $\frac{1}{7}$ to get $8 \times 8\frac{2}{7} = \frac{56}{7} \times \frac{58}{7} = \frac{3248}{49} = 66\frac{14}{49}$.

The square of the fraction is $(\frac{1}{7})^2 = \frac{1}{49}$. Putting this all together we get $66\frac{14}{49} + \frac{1}{49} = 66\frac{15}{49}$.

If you use the multiplication of alternating numbers tip 56×58 is the same as 57^2 minus 1, and to solve 57^2 you could use the tip for squaring two-digit numbers:

Step 1: $7^2 = 49$, rightmost digit is 9 carry 4

Step 2: 2(5 x 7) = 70 + 4 = 74, second digit is 4, carry 7

Step 3: 5^2 = 25 + 7 = 32

Put 32, 4, and 9 together to get the answer 3,249 and subtract 1 using the multiplication of alternating numbers tip for the answer 3,248. Then it's just a matter of dividing 3,248 by 49 to get 66 with a remainder of 14. The square of the fraction is $1/49$ and when we add it to $14/49$ we get $15/49$.

The normal way is a little tougher:

$$8\tfrac{1}{7} \times 8\tfrac{1}{7} = \tfrac{57}{7} \times \tfrac{57}{7} = \tfrac{3249}{49} = 66\tfrac{15}{49}$$

DID YOU KNOW?

The oldest know mathematics texts are the Babylonian tablet *Plimpton 322* circa 1900 BC, the Egyptian *Rhind Mathematical Papyrus* circa 2000–1800 BC and the Egyptian *Moscow Mathematical Papyrus* circa 1890 BC. All these texts mention the Pythagorean Theorem which seems to be the next mathematical development after basic arithmetic and geometry. The study of mathematics begins in the 6th century BC with the Pythagoreans, who coined the term "mathematics" from the ancient Greek *mathema*, meaning "subject of instruction".

How to Find Square Roots Without a Calculator

The square root of a number is a value that when multiplied by itself yields the number, for example 2 x 2 = 4 or 2^2 = 4, so the square root of 4 is 2. This is usually represented as $\sqrt{4}$ but can also be represented as 4 raised to the ½ power or $4^{½}$. When most people are asked to find the square root of a number they go running for a calculator (or just take off running). In a few simple steps and just using basic math we'll show you how to find a square root without a calculator, no running necessary.

Let's say we need to find the square root of 66,564, $\sqrt{66,564}$ or $66,564^{½}$. Since this is a five-digit number we start by finding the two hundreds our number lies between:

$$100^2 = 10,000$$
$$200^2 = 40,000$$
$$300^2 = 90,000$$

We can stop right there, 66,564 is between 40,000 and 90,000 so the first number of the square root is a 2.

To find the second digit subtract 40,000 from 66,564 and divide by 2 x 200:

```
   66,564                           60
 - 40,000      2 x 200 = 400   400)26,564
   26,564                         24,000
```

To test if 6 is the 2nd number of our square root we'll do a test:

$$400 + 60 = 460, 460 \times 60 = 27,600$$

27,600 is greater than our dividend of 26,564 so the 2nd digit isn't a 6. This time let's try 50:

$$400 + 50 = 450, 450 \times 50 = 22,500$$

The product is smaller so the second number of our square root is 5! Now it's just a matter of repeating how we find the second number to get the next digit:

```
  26,564                              ____8
 - 22,500        2(200 + 50) = 500    500)4,064
   4,064                                  4,000
                                             64
```

To test if 8 is the number of our square root we'll do a test:

$$500 + 8 = 508, 508 \times 8 = 4,064$$

There's no remainder so the 3rd digit is 8, and our square root is 258. To test if we're right we'll square 258: $258^2 = 66,564$

So $\sqrt{66,564}$ = 258, no calculator needed.

Another Way to Find Square Roots

In this method we are going to break a number up into smaller groups to find the square root. If we wanted to find the square root of 427,716 we would break it up into 3 groups of two-digits, 42, 77, and 16.

Step 1: Working with the first group of numbers, 42, we need to find a one-digit number whose square does not exceed the number. $6^2 = 36$ and $7^2 = 49$ so it looks like our first number is 6.

Step 2: Subtract 6^2 from 42:
$$\begin{array}{r} 42 \\ -36 \\ \hline 6 \end{array}$$

Step 3: Add the second two-digit number, 77, to the number we just found 6, and divide 677 by two times the first digit:

$$\begin{array}{r} 5 \\ 125{\overline{\smash{)}677}} \\ 625 \\ \hline 52 \end{array}$$

The second digit is 5.

We need a 3-digit divisor so we can get a 1-digit quotient so divide 67 by 12 to get 5 and add 5 to the end of the divisor

Step 4: Bring down the last two digits, 16, and put them by the remainder of the last step, 52, to make 5,216.

Step 5: Divide 5,2165 by two times the first and second digits of the answer, 2 x 65 = 130:

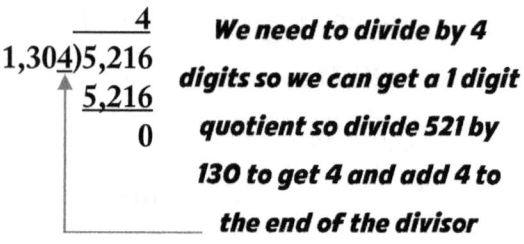

1,304 goes into 5,216 exactly 4 times so our 4th digit is a 4 and our square root is 654. Let's check to see if we're right:

654^2 = 427,716, so our answer checks good!

DID YOU KNOW?

Amicable numbers are two different numbers whose proper divisors' sum equals the other number. The smallest pair of amicable numbers is 220 and 284. They are amicable because the proper divisors of 220 are 1, 2, 4, 5, 10, 11, 20, 22, 44, 55 and 110, and those numbers added together equal 284, and the proper divisors of 284 are 1, 2, 4, 71 and 142, and they add up to 220. Unlike perfect numbers there are millions of amicable pairs and as of April 2017 there are over 1,200,000,000 known pairs.

You Know More Latin Than You Think!

You've been speaking Latin and didn't know it! Sixty percent of the English language comes from Latin, check out a few of those words you've been using unknowingly:

Plus is Latin for "more", and minus means "less".

The number to be subtracted from is the minuend which is Latin for "to be diminished", and the number which is subtracted is called the subtrahend which means "to be subtracted".

In multiplication the upper quantity is called the multiplicand which is Latin for "to be multiplied". The lower quantity is called the multiplier and the result is the product.

In division the number above the bar (e.g. $\frac{2}{4}$) is the dividend which is Latin for "to be divided", the lower number is the divisor. The result of division is the quotient which is Latin for "how often".

"Post" is Latin for after, "ante" means before.

"Bin" as in binary comes from the Latin word "bini".

"Cent" as in centennial comes from the Latin word "centum" which means one hundred.

Prime Numbers Tips

Prime numbers are special numbers indivisible by any other number except 1. For example, 53 is a prime number because it can only be divided by 1. You can impress people with your knowledge of prime numbers by using the following 5 rules:

1. The number 2 is the only even prime number.
2. No prime number can end in 5 except 5.
3. After the number 10 prime numbers can only end in 1, 3, 7, or 9.
4. After the number 2 and 3 if a prime number is increased or decreased by 1 one of the results will be divisible by 6.
5. The product of two prime numbers can never be a square number.

With these five rules you can tell if just about any number is prime.

DID YOU KNOW?

Fibonacci numbers were named after Leonardo Fibonacci but he actually learned about them from Arab mathematicians who learned about them from Hindu mathematicians who had known about them since the early sixth century.

Math Tricks

There are 28 Math Tricks ranging from instant mental math tricks, tricks to make you seem like you have psychic abilities, and my favorite math-based card tricks. Find tricks you like and try them out on your friends and family. Performing these math tricks is a great way to build your confidence and reduce your anxiety of doing math in public. And if you can explain how the tricks work to your audience you will have fully grasped and understood the math behind them.

Every trick has a detailed explanation of how and why they work along with some tips on how to perform them. Some you will be able to do right away and others may take a little practice to do them correctly. Keep working on them and you'll be entertaining audiences in no time.

The 1,089 Math Trick

I've seen this trick done many times online and found many references to it in old math books dating back to the 1800's but no one ever bothers to explain *why* the trick works. If a magician pulls a rabbit out of his hat you would expect him to know where it came from, right?

This is how to do the trick: ask someone to think of any three-digit number whose first and last digits are at least two numbers apart, for example 345 is okay but 344 and 343 are not. Reverse this number, or find its reciprocal, and subtract the smaller number from the larger one. Have them add the reciprocal of the answer to this number but before they can give you the sum you pull out a piece of paper with the answer, 1,089, already written on it. They then look at you with amazement and ask, "What sorcery is this?" Tell them no sorcery was involved, just a little algebra. Give them the following explanation for how the trick works:

Any three-digit number can be represented as:

$$100a + 10b + c$$

The reciprocal of this can be represented as:

$$100c + 10b + a$$

Subtracting the two expressions can then be represented as:

$$100a + 10b + c - (a + 10b + 100c)$$

$$100a + 10b + c - a - 10b - 100c$$

$$\begin{array}{r} 100a + 10b + c \\ -\ (a\ \ +\ 10b\ +\ 100c) \\ \hline 99a\ \ \ \ \ \ \ \ \ \ -\ 99c \\ 99(a-c) \end{array}$$

The difference between *a* and *c* will always be a number between 2 and 9 and as you can see in the table below there are only 8 possible results of 99(a - c). You can also see why the first and last numbers must be at least 2 or more numbers apart, if a - c resulted in a 1 or 0 you would only get one answer, either 99 or 0.

a - c			
2	1	9	8
3	2	9	7
4	3	9	6
5	4	9	5
6	5	9	4
7	6	9	3
8	7	9	2
9	8	9	1

Adding the reciprocal of any of these numbers will always give you 1,089. Using a little algebra again will show you why this is so. Since we know every 3-digit number can be represented as 100a + 10b + c adding the reciprocal would look like this:

$$
\begin{array}{rrrr}
 & 100a + & 10b + & c \\
+ & a + & 10b + & 100c \\
\hline
 & 101a + & 20b + & 101c
\end{array}
$$

The expression 99(a - c) always results in a number with 9 as the middle number so the value for *b* will always be

$$20b = 20(9) = 180$$

The values for *a* and *c* are always multiplied by 101, and as you can see in the table above the sum of *a* and *c* is always 9. This means no matter which number and reciprocal you're adding you will always get the same answer. For example, if the number selected results in 198, *a* and *c* are 1 and 8 respectively:

$$101 \times 1 = 101$$
$$101 \times 8 = 808$$
$$808 + 101 = 909$$
$$909 + 180 = 1,089$$

If the number selected results in 297 then *a* and *c* are 2 and 7:

$$101 \times 2 = 202$$
$$101 \times 7 = 707$$

$$202 + 707 = 909$$
$$909 \text{ and } 180 = 1{,}089$$

The reason a and c always equal 9 is because we created a multiple of 9 when we multiplied a and c by 99. When you add the digits of a multiple of 9 together they always equal 9. You also create a multiple of 9 by subtracting the reciprocal of a number from itself. Let's try 123,456:

$$654{,}321 - 123{,}456 = 530{,}865$$
$$5+3+8+6+5 = 27$$
$$2 + 7 = 9$$

DID YOU KNOW?

A perfect number is a positive integer equal to the sum of its divisors excluding the number itself (also known as its aliquot sum). Also, a perfect number is a number half the sum of all of its positive divisors (including itself). The first perfect number is 6 because its divisors are 1, 2, and 3, and $1 + 2 + 3 = 6$. The number 6 is also equal to half the sum of all its positive divisors: $(1 + 2 + 3 + 6) / 2 = 6$. The next perfect number is 28, $28 = 1 + 2 + 4 + 7 + 14$ and $1 + 2 + 4 + 7 + 14 + 28/2 = 28$. This is followed by the perfect numbers 496 and 8,128. It isn't known whether there are any odd perfect numbers. As of 2016 there are only 49 known perfect numbers.

The Number Thought Upon Trick

This is the first of several "number thought upon" tricks. Number thought upon tricks have been popular math tricks for hundreds of years, I've found references to them in math books dating all the way back to the 1600's. This one uses part of the trick we just learned about subtracting the reciprocal from a number always resulting in a multiple of 9.

Ask someone to come up with a really long number, the longer the better. Then have them subtract the reciprocal of the number but don't give you the answer. First ask them to pick their favorite (or least favorite) non-zero number from the answer and strike it out. Have them add the remaining digits together and tell you the sum. You will amazingly be able to tell them exactly what number they struck out. How did you do it?

Let's say they used the following number:

```
   5 7, 8 6 3, 4 2 1, 0 4 9, 3 2 1
 - 1 2, 3 9 4, 0 1 2, 4 3 6, 8 7 5
   ─────────────────────────────────
   4 5, 4 6 9, 4 0 8, 6 1 2, 4 4 6̸
```

You can see they struck out the last 6, and the sum of the remaining digits is:

$$4 + 5 + 4 + 6 + 9 + 4 + 8 + 6 + 1 + 2 + 4 + 4 = 57$$

You instantly know the number they struck out was a 6 because the next multiple of 9 is 63, and 63 - 57 = 6. When we subtracted the reciprocal of the number we created a multiple of 9, and as you'll recall from the previous trick the digit sum of a multiple of 9 always equals a multiple of 9, and if you keep adding digits together you will always get 9.

MATH FUNNIES

What was the problem when the math teacher couldn't see straight?

The Fibonacci Trick

This trick is based on the Fibonacci sequence, a series of numbers where a number is found by adding up the two numbers before it. Starting with 0 and 1 the sequence goes 0, 1, 1, 2, 3, 5, 8, 13, 21, 34, and so on. For this trick though you can ask someone to pick any two single digit numbers they like for their Fibonacci sequence. Have them keep adding numbers until they have 10 terms. Ask them to find the sum of all the numbers in their sequence, but much to their surprise you will know the answer before they can punch all the numbers into their calculator! How did you pull that off?

This is how the trick works: let's say they picked the numbers 2 and 4, then their first term would be 2, the second 4, the third 6, etc. If we call the first term x and the second term y the sequence can be represented algebraically as shown on the next page.

2	x
4	y
6	x + y
10	x + 2y
16	2x + 3y
26	3x + 5y
42	5x + 8y
68	8x + 13y
110	13x + 21y
178	21x + 34y
462	55x + 88y

Both 55 and 88 are divisible by 11 so we can reduce the expression to 11(5x + 8y). Where have we seen that before? That's right, it's the algebraic expression for the 7th term. So now we've figured out that little tidbit of information we can use it to our advantage: the 7th term multiplied by 11 always equals the sum of all the terms. The 7th term is 42, and 42 x 11 = 462.

If you were paying attention in the Math Tips section of the book you'll recall there is an easy way to multiply any number by 11 - first write down the 1st digit of the number, then add each consecutive number until you get to the last number. In this case to multiply 42 x 11 start with a 4, add 4 + 2 to get 6 for the next number, and then write down the last number, 2.

Let's do a more difficult example with 7 and 13 as the starting numbers for our sequence:

$$\begin{array}{r} 7 \\ 13 \\ 20 \\ 33 \\ 53 \\ 86 \\ 139 \\ 225 \\ 364 \\ \underline{589} \end{array}$$

Counting four up from the bottom we see the 7th term is 139. Using our tip for multiplying by 11 our first digit is going to be 1, the second number 1 + 3 = 4, but 9 + 3 = 12 so mentally carry a 1 and add it to the 4 so our number is now 152, adding in the last digit gives us 1,529, which is the sum of all the numbers in the sequence.

Fibonacci's number sequence 0, 1, 1, 2, 3, 5, 8 … can be used to make a series of boxes which when connected with curves closely approximate (but not exactly) the Golden Spiral, a logarithmic spiral whose growth factor is represented by the Greek letter Phi φ, the golden ratio 1.6108. The Golden Spiral is found everywhere in nature, giving nautilus and snail shells their characteristic shape, perfectly spacing the seeds in sunflowers, and even in the shape of some spiral galaxies.

To make Fibonacci boxes start by making two boxes with sides of 1, place a box with sides of 2 on top, then a box with sides of 3 beside those three boxes, and continue drawing boxes in this manner until you have ten boxes.

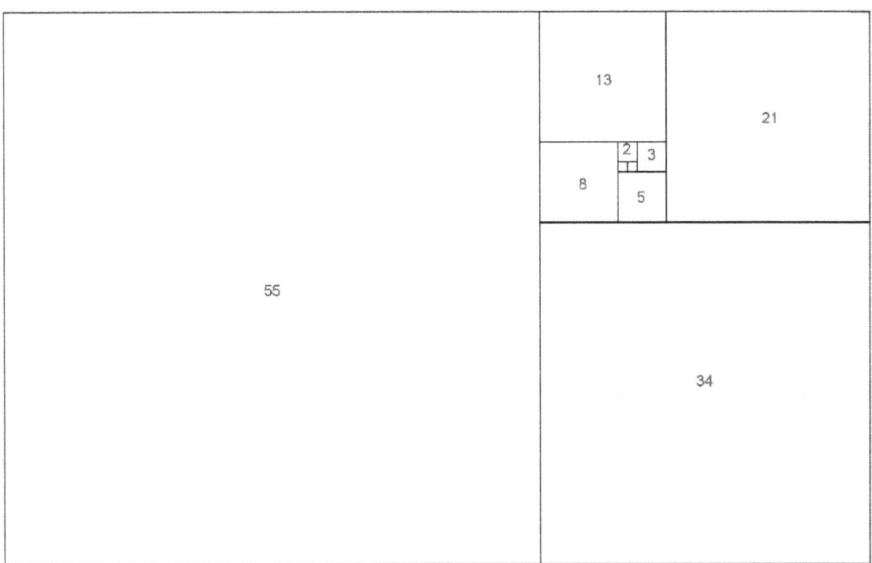

If you draw a quarter curve in each box and connect each box until you reach the second box with sides of 1 you'll have a pretty good approximation of the Golden Spiral. As you draw more and more boxes you'll get closer and closer to φ, 1.61803, but if you stop at ten you'll have a large box with sides of 55 beside a smaller box with sides of 34 which have a ratio of 1.61764. The percent difference between the two is .024% which is close enough for a good spiral.

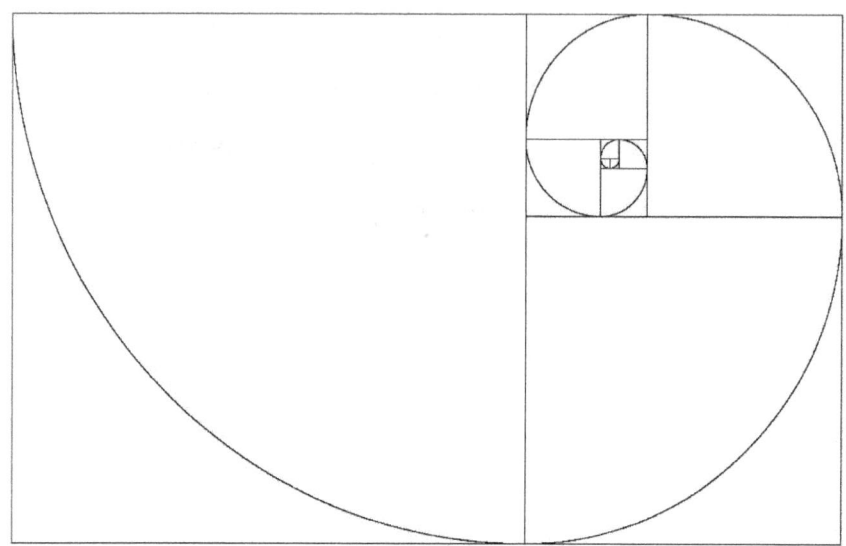

The Golden Spiral can be found all over nature and the Universe, here are just a few examples.

Cephalopod Ammonite Fossil from the Devonian Period, ~350-400 million years old

A modern day snail shell

Spiral Nebula, from H.G. Wells "A Short History of the World"

The 99 Math Trick

With this trick you secretly write down a number and ask someone to pick a number of their own and have them do a little math. The answer they get always ends up being your number regardless of what number they choose!

First, choose a number between 10 and 49, let's say 37. Don't tell the other person this number.

Subtract the number from 99: 99 - 37 = 62 and tell the person your number is 62. Keep the number 37 to yourself, for added drama write it on a piece of paper and hide it in your pocket.

Ask them to pick a number between 50 and 99, say 42, and add it to your number: 62 + 42 = 104.

Cross out the hundreds digit and add it to the ones digit: ~~1~~04 + 1 = 5.

Subtract 5 from their number, 42, and the answer is your secret number, 37! You tell them, "Where have I seen that number before? Oh yeah, in my pocket!" and to their astonishment you pull out a piece of paper with the number 37 written on it. How does this work?

The number you chose between 10 and 49 can be represented algebraically using x for the tens digit and y for the ones digit,

$10x + y$. Adding 62, your secret number subtracted from 99, can be represented as $10x + y + 62$. Crossing out the hundreds digit is the same as subtracting 100, and adding the 1 back in makes the expression:

$$10x + y + 62 - 100 + 1 \text{ or}$$
$$10x + y - 37$$

Subtracting their original number can be represented as

$$(10x + y - 37) - (10x + y)$$
$$10x + y - 37 - 10x - y = 37$$

No matter what numbers you use as long as you keep them in the right range of numbers, 10 to 49 and 50 to 99 (to make sure the hundreds place is always a 1), you'll end up with whatever number you choose.

MATH FUNNIES

Math - the only subject that counts.

The Guess the Number Trick

Ask someone to pick two different numbers (can't use the same number twice) and subtract the smaller combination of the numbers from the larger. For example, if they chose 5 and 2 they would subtract 25 from 52 to get 27. Ask them what their answer was and tell them if they will share with you the smaller of the two original numbers you can tell them the other number. In the example above you would instantly be able to tell them "Your other number was a 5". How does this work?

All you do is divide their number by 9 and add the number they gave you. $27 \div 9 = 3$, and $3 + 2 = 5$. This works because whenever you subtract a reciprocal from a number you create a multiple of 9. This can be represented as:

$$\begin{array}{r} 10a + b \\ -(a + 10b) \\ \hline 9a - 9b \\ 9(a-b) \end{array}$$

Divide by 9 and you're just left with $(a - b)$. When you add the smaller number, b, you are left with a.

$$9(a - b) / 9 = (a - b)$$
$$(a - b) + b = a$$

The Instant Addition Trick

This trick is another play on the multiples of 9 theme and a demonstration of the magic of the number 9. Bet someone you can instantly sum any 3 large numbers they can imagine, and to make it more interesting you will add 2 equally long numbers to their 3. Before they can fire up a calculator app on their phone you will have found the sum of the numbers.

Let's say they pick the numbers 739,187,259, 349,747,457, and 988,360,327. Your 2 numbers are 260,812,740 and 650,252,542. As soon as they write down their last number you immediately start writing the answer, 2,988,360,325. How did you do it so fast?

The secret is to add 2 to the front of their last number and subtract 2 from the last number to get the answer. This works because you slipped your two numbers in between their numbers and made sure your numbers make their numbers add up to 9. Your numbers are the second and the fourth, and the sum of the first two numbers and the second two numbers is 999,999,999. This means their last number is being added to 999,999,999 + 999,999,999 which is 1,999,999,998, and the answer is always going to be the last number they write down with a 2 added to the left and 2 subtracted from the number on the right.

```
  739,187,259
  260,812,740
  349,747,457
  650,252,542
  988,360,327
2,988,360,325
```

By making sure your numbers make their numbers add up to 9 whatever number they pick last will always be added to 1,999,999,998.

```
1,999,999,998
  988,360,327
2,988,360,325
```

If you want to add two more numbers just add a 3 to the front and subtract 3 from the last digit, add two more digits and add a 4 to the front and subtract 4 from the last digit, etc.. Using this method you can take this as far as you want, if you wanted to do a column of 21 numbers you would just need to remember to add a 10 to the front and subtract 10 from the last digit.

MATH FUNNIES

You know what seems odd to me? Numbers that aren't divisible by 2.

The Dice Trick

This is a very old trick dating all the way back to the 1600's. Ask someone to roll two dice without allowing you to see the results and then ask them to do the following:

- Multiply the number of the first dice by 5
- Add the number 6,
- Multiply the sum by 2,
- And add the number from the second dice.

When they tell you their answer in a flash you will be able to tell them what their dice rolls were in order!

The secret to this is to subtract the number 12 from whatever number they give you, the first number will always be the roll of the first die and the second number will tell you the roll of the second die. How does this work?

Let's call the roll of their first dice a and the roll of their second dice b. When you ask them to multiply their first die by 5 you get 5a, adding 6 results in 5a + 6, and multiplying that by 2 gives you 2(5a + 6). Adding the result of the second die gives you 2(5a + 6) + b and simplifying the expression yields 10a +12 + b. So no matter what they roll all you do is subtract 12 and you will have the answer:

$$10a + 12 + b - 12 = 10a + b$$

I've found many variations of this trick in very old math books, some with two dice some with three, some with different formulas, etc. In my version, I used the number 12 but that doesn't mean you can't come up with your own version of this trick. Just play around with different expressions and if you come up with something you think is good share it with us!

MATH FUNNIES

Why was the angle freezing?

The Day of the Week Trick

This trick will show you how to calculate the day of the week for any historical event (like the day you were born!). To calculate the week day of any historical event all you need to know is the "Julian date", the number of how many days into the calendar year your number is, and this formula:

$$S = Y + D + \left(\frac{Y-1}{4}\right) - \left(\frac{Y-1}{100}\right) + \left(\frac{Y-1}{400}\right)$$

S is the day of the week, Y is the year, and D is the Julian date. Some calendars give you the Julian date as the running total of the number of days remaining in the year, which you would just subtract 365 to get your number. If you don't have this kind of calendar handy you can use this table to calculate the Julian date:

January: 31 days: 31
February: 28 days: 59
March: 31 days: 90
April: 30 days: 120
May: 31 days: 151
June: 30 days: 181
July: 31 days: 212
August: 31 days: 242
September: 30 days: 273
October: 31 days: 304
November: 30 days: 334
December: 31 days: 365

Or you can use this simple rhyme I learned as a child which has been used since the 1400's:

"Thirty days has September, April, June and November. All the rest have 31 except February, it's a different one. It has 28 days clear and 29 in a leap year."

Let's calculate what day of the week we first landed on the moon, July 20, 1969. To find the Julian date we add up the days through the end of June and add the 20 days in July for a Julian date of 201 and plug these into the formula.

$$1969 + 201 + \left(\frac{1969-1}{4}\right) - \left(\frac{1969-1}{100}\right) + \left(\frac{1969-1}{400}\right) = 2{,}647$$

We round the fractions to the nearest whole number and get 2,647. To find the day of the week we need to divide our value for S by 7 and the remainder gives us the day of the week.

```
    378
7)2,647
   21
   54
   49
    57
    56
     1
```

The remainder of 1 gives us the day of the week:

0 = Saturday
1 = Sunday
2 = Monday
3 = Tuesday
4 = Wednesday
5 = Thursday
6 = Friday

The remainder of 1 tells us Apollo 11 landed on the moon on a Sunday.

Why did we need to add and subtract all those fractions? It's because of those pesky leap years. It takes the Earth approximately 365.242189 days, or 365 days, 5 hours, 48 minutes and 45 seconds, to orbit the Sun. This is called a Tropical Year. Since the Gregorian Calendar we've used since 1582 only has 365 days if we didn't add a leap day to the shortest month just about every 4 years we would lose almost 6 hours off our calendar every year. This doesn't sound like much but after about 100 years it would add up to 24 days! When you are trying to calculate a particular day of the week you will need to know if the year you are using is a leap year. You could go to the interwebs and look it up or you just do this simple test: if you can divide the year by 4 it's a leap year <u>unless</u> the year can be evenly divided by 100 in which case it is <u>not</u> a leap year unless the year is evenly divisible by 400 in which case it <u>is</u> a leap year. This means the years 2000 and 2400 are leap years but 1800, 1900, 2100, 2300 and 2500 are not leap years. If your year is a leap year and your date falls after February 28 you'll need to add an extra day to your Julian day. Let's see this in action:

On what day did our founding fathers sign the Declaration of Independence?

1776 is an even year so let's see if it's divisible by 4:

```
   444
4)1776
   16
   17
   16
   16
   16
    0
```

1776 is divisible by 4 and not evenly divisible by 100 or 400 so it was a leap year. This means we'll need to add a day to the Julian date for July 4th. There are 181 days through June plus the 4 days in July and the 1 leap day gives us:
181 + 4 + 1 = 186

$$1776 + 186 + \left(\frac{1776-1}{4}\right) - \left(\frac{1776-1}{100}\right) + \left(\frac{1776-1}{400}\right) = 2{,}392$$

Adjusts for leap years **Removes years divisible by 100** **Adds back years divisible by 400**

```
    341
7)2,392
     21
     29
     28
     12
      7
      5
```

The remainder of 5 gives us the day of the week:

0 = Saturday
1 = Sunday
2 = Monday
3 = Tuesday
4 = Wednesday
5 = Thursday
6 = Friday

Using a little math we now know the Declaration of Independence was signed on a Thursday.

Another Day of the Week Trick

This method for calculating the day of the week an event occurred is attributed to Oxford professor of mathematics Charles Lutwidge Dodgson, who you might also know as Lewis Carroll, the author of "Alice's Adventures in Wonderland". Let's find out what day the 4th of July was in 1976:

Step 1: Take the last two digits of the year and divide by 12: 76 ÷ 12 = 6 remainder 4

Step 2: Add the remainder to the quotient: 6 + 4 = 10

Step 3: Divide the remainder in Step 1 by 4: 4 ÷ 4 = 1 (if there is a remainder here just ignore it)

Step 4: Add the quotient in Step 3 to the sum in Step 2: 1 + 10 = 11

Step 5: Add the day of the month to the sum in Step 4: 4 + 11 = 15

Step 6: Using the following list add the month's number to the sum in Step 5. If the year is a leap year and the month is January or February you'll need to subtract 1 from the month's number:

January = 1	May = 2	September = 6
February = 4	June = 5	October = 1
March = 4	July = 0	November = 4
April = 0	August = 3	December = 6

So 15 + 0 for July is 15.

Step 7: If the year is in the 19th century add 2, 2000 or later subtract 1, between 1900 and 1999 do nothing.

Step 8: Divide the sum in Step 7 by 7 and the remainder tells you the weekday:

        0 = Saturday
        1 = Sunday
        2 = Monday
        3 = Tuesday
        4 = Wednesday
        5 = Thursday
        6 = Friday

$15 \div 7 = 2$ remainder 1, and a quick check verifies July 4th, 1976 was a Sunday!

MATH FUNNIES

An opinion without π is just an onion.

Another Number Thought Upon Trick

There are many kinds of number thought upon tricks, this is my version of one I found in an old book from the 1600's. With this trick you:

- ask someone to pick a number,
- add 1 to the number,
- square both numbers, and then
- subtract the smaller of the two numbers.

When they give you the difference between the numbers you will be able to tell them their original number. How does this work?

If we call their number n adding 1 to their number could be represented as n + 1.

When we square n we get n^2, squaring n + 1 is

$$(n + 1)^2$$
$$(n + 1)(n + 1)$$
$$n^2 + 2n + 1$$

The first number squared will always be the smaller of the two numbers so subtracting it gives you:

$$n^2 + 2n + 1 - n^2$$
$$2n + 1$$

Now all you have to do is subtract 1 and divide by 2 to get their original number,

$$2n + 1 - 1 = 2n$$
$$2n \div 2 = n$$

Let's say their number was 37, so 37 + 1 = 38.

$$37^2 = 1,369$$
$$38^2 = 1,444$$
$$1,444 - 1,369 = 75$$

They tell you their number is 75 and mentally you subtract 1 to get 74, and half of 74 is 37, their original number.

If you have trouble dividing a number by 2 round down to the next number with a zero and find half of that number and half of the difference. For example, if we had trouble with the number 74 we would round it down to 70, half of 70 is 35, and half the amount we rounded down, 4, is 2. Then 35 + 2 = 37. What if it's an odd number, you say? Won't happen, by subtracting the squares and subtracting 1 you always end up with an even number.

The Number 34 Trick

For this trick make a 4 x 4 table with all the numbers from 1 to 16 in sequential order. Then have someone pick a number from each column and cross out the numbers above, below, and beside the number. Ask them to find the sum of the numbers they chose but before they can answer you tell them the sum of their numbers is 34! How did you know that?

1	2	3	4
5	6	7	8
9	10	11	12
13	14	15	16

This is what the table should look like if someone picks the number 5 in the first column.

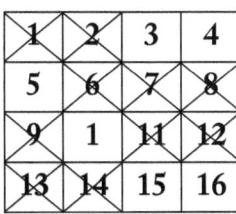

If they picked the number 10 from the 2nd column it would look like this.

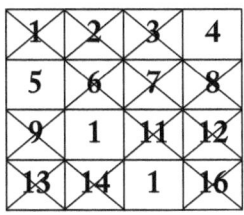

If they picked the numbers 5, 10, 15, and 4 the sum of the four numbers is 34. How did you know this beforehand?

					Σ:
	1	2	3	4	10
	5	6	7	8	26
	9	10	11	12	42
	13	14	15	16	58
Σ:	28	32	36	40	

The secret is 34 is the average of the sums of each row and column.

$10 + 26 + 42 + 58 = 136$

$28 + 32 + 36 + 40 = 136$

$136 + 136 = 272$

$272 \div 8 = 34$

No matter what numbers they pick if only one number is used per row and column the sum will always be 34. If you do this for a 5x5 table the number would be 65, for a 6x6 table 111, a 7x7 table would be 175, and so on for any size you want if you do everything the same as we did with the 4x4 table.

Now let's learn a variation on this trick with a bigger table and different numbers.

MATH FUNNIES

This is what the mathematician gave his girlfriend:

$9x - 7i > 3(3x - 7u)$

(If you don't get it the answer's in the back!)

The Addition Table Trick

This trick is like the Number 34 Trick but we can use any numbers we want for the table.

69	91	92	71	89	83	49
39	61	62	41	59	53	19
45	67	68	47	65	59	25
47	69	70	49	67	61	27
28	50	51	30	48	42	8
31	53	54	33	51	45	11
34	56	57	36	54	48	14

This 7x7 table is filled with seemingly random numbers.

69	91	92	71	89	83	49
39	61	62	41	59	53	19
45	67	68	47	65	59	25
47	69	70	49	67	61	27
28	50	51	30	48	42	8
31	53	54	33	51	45	11
34	56	57	36	54	48	14

With 28 selected as the 1st number, cross out everything above, below, and beside the number.

69	91	92	71	89	83	49
39	61	62	41	59	53	19
45	67	68	47	65	59	25
47	69	70	49	67	61	27
28	50	51	30	48	42	8
31	53	54	33	51	45	11
34	56	57	36	54	48	14

The rest of the numbers are 28, 69, 62, 33, 89, 59, and 14. We already know the sum is 354, but let's add them up just to be sure.

$$28 + 69 + 62 + 33 + 89 + 59 + 14 = 354$$

How did we know that? Is 354 the average of the sums of the columns and rows like the last trick? Nope, this time we made a big addition table and 354 is the sum of the 14 numbers of the addition table.

Σ	26	48	49	28	46	40	6
43	69	91	92	71	89	83	49
13	39	61	62	41	59	53	19
19	45	67	68	47	65	59	25
21	47	69	70	49	67	61	27
2	28	50	51	30	48	42	8
5	31	53	54	33	51	45	11
8	34	56	57	36	54	48	14

Every number in the table is the sum of the numbers at the top and side of the table. The first cell has a 69 which is 43 + 26.

The cell next to it has a 91, and 43 + 48 = 91. The middle cell has a 49, and 21 + 28 = 49, see how it works? If we add the addition table numbers together we get 354.

43+13+19+21+2+5+8+26+48+49+28+46+40+6 = 354

Any combination of numbers selected is going to add up to 354 if only one number is picked per column and row. I used a random number generator for the addition table numbers and a spreadsheet to quickly fill the cells. You can make your own addition table trick using any size table or combination of addition table numbers.

If you want to quickly fill the table in using a spreadsheet make the cell value for the addition table number from the left side an absolute value, like A2, and reference the other addition table value normally, so your cell formula would look like =A2+B1 if you were starting in the upper left-hand corner of the spreadsheet. Do this for each cell on the left side of the table and all you have to do is copy the left cell and paste into the other cells in the row. This way when you copy and paste the reference for the left cell won't change but the reference for the column addition table numbers will change.

Yet Another Number Thought Upon Trick

This is the last one, I promise, and it has an interesting twist at the end you should appreciate. Here's how to do the trick:

- Have someone pick any number and multiply it by 10.
- Then pick any number between 1 and 9 and multiply it by 9.
- Subtract the product from the other number.

When they give you the answer you will know both of their numbers. If we call the first number m and the second number n we can represent what they did as:

$$10m - 9n = x$$

To figure out their numbers all you have to do is mentally take the number in the ones place and add it to the other two numbers, the sum is their first number and the number in the ones place is their second number. Let's say their first number is 46 and their second number is a 5:

$$10m - 9n = x$$
$$10(46) - 9(5) = 415$$

Take the 5, which is their 2nd number, and add it to 41 to get 46, their 1st number. Why does this work?

Algebraically *x* can be represented as:

$$x = 100a + 10b + c$$

When we move the ones digit and add it to the other two digits it's the same as subtracting *c* and dividing by 10, and then adding *c* back in again:

$$x = \frac{100a + 10b + c - c}{10} + c$$

$$x = 10a + b + c$$

Substituting our value for *x* into our new formula we get

$$x = 10(4) + 1 + 5$$

$$\text{So } x = m = 46$$

Their other value, *n*, is always equal to the value for *c*, in this case 5.

MATH FUNNIES

---

plan

$$(p + l)(a + n) = pa + pn + la + ln$$

I just foiled your plan.

The Multiplication Summation Trick

In this trick you'll ask someone to pick a number of 3, 4, 5, or more digits (it's up to you) and tell them if they will give you another number with the same number of terms you will simultaneously do two multiplications and add their products together in your head! This is how it works: let's say they pick the number 8,375 for their multiplicand and 2,135 for their multiplier. You write down 8,375 twice, put their multiplier under the left number and tell them you will pick the second multiplier, which in this case is 7,864.

$$\begin{array}{cc} 8{,}375 & 8{,}375 \\ \times\,2{,}135 & \times\,7{,}864 \end{array}$$

Without hesitation you work out the answer to be 83,741,625. They don't believe you could do all that in your head and grab a calculator to check your answer:

$$\begin{array}{ccc} 8{,}375 & 8{,}375 & 17{,}880{,}625 \\ \times\quad 2{,}135 & \times\quad 7{,}864 & +\,65{,}861{,}000 \\ \hline 17{,}880{,}625 & 65{,}861{,}000 & 83{,}741{,}625 \end{array}$$

How does this work? The trick is when you choose your multiplier make sure every number you pick makes their corresponding number equal 9. The first digit of their multiplier is a 2 so you make yours a 7, 2 + 7 = 9. Their second number is a 1 so your second number is 8, and so on. Now if you add the numbers of the multipliers together you get 9,999:

$$2{,}135 + 7{,}864 = 9{,}999$$

To quickly come up with the answer write down their multiplicand minus 1 and then the numbers making each of those digits equal 9:

$$8{,}375 - 1 = 8{,}374$$
$$8 + 1 = 9, 5^{th} \text{ digit is a } 1$$
$$3 + 6 = 9, 6^{th} \text{ digit is a } 6$$
$$7 + 2 = 9, 7^{th} \text{ digit is a } 2$$
$$4 + 5 = 9, 8^{th} \text{ digit is a } 5$$

Making your multiplier out of numbers which make their multiplier add up to 9 is the same as if we multiplied their multiplicand by 9,999:

$$\begin{array}{r} 8{,}375 \\ \times 9{,}999 \\ \hline 83{,}741{,}625 \end{array}$$

Multiplying a number of any size by the same number of 9's always gives you this result, for example:

$$\begin{array}{r}321\\ \times999\\ \hline 320{,}679\end{array}$$

The first 3 digits are the multiplicand minus 1:

$$321 - 1 = 320$$

And the next three digits makes those numbers add up to 9:

$3 + 6 = 9$, 4th digit is 6,
$2 + 7 = 9$, 5th digit is 7,
$0 + 9 = 9$, 8th digit is 9.

Telling them you are going to simultaneously do two multiplications and add the products together makes this sound more complicated. If you take any multiplication problem, split the multiplier up into two numbers and use the same multiplicand you get the same result, for example:

$$\begin{array}{r}111\\ \times36\\ \hline 3{,}996\end{array}$$

If we split the multiplier 36 up into 24 and 12, multiply them by 111 and add their products you get the same answer:

111	111	2,664
x 24	x 12	+ 1,332
2,66	1,332	3,996

The 1969 Trick

This trick uses the awesome year 1969 to help you seem to psychically predict a number. This is how to do the trick:

- Ask someone to pick a number between 100 and 1,000
- Have them add their number to 1969
- Tell them to strike out the number in the thousands place and add it to the remaining number
- Then they subtract the result from their original number

Before they can finish punching the numbers into their calculator you tell them answer is 29. How could you have possibly known the answer?

Let's say they picked the number 42, it gets added to 1969 and the thousands number gets struck out and added to the remaining numbers:

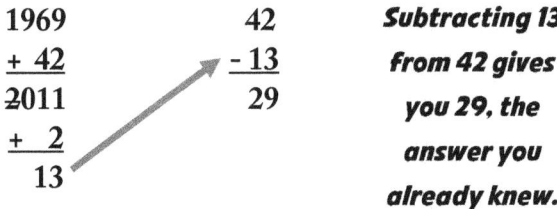

Subtracting 13 from 42 gives you 29, the answer you already knew.

If we represent this algebraically you'll see why it works. We'll call their number n, and striking out the number in the thousands place and adding it to the remainder of the number is the same as subtracting 2,000 and adding back a 2:

$$((1969 + n) - 2000 + 2) - n$$

We're adding and subtracting *n* so it cancels out leaving 1969 – 2000 + 2 or 1969 – 1998 which is 29. This is always the answer if they pick a number between 100 and 1,000. Actually, they need to pick a number between 31 and 1,030 but that might tip them off to how the trick works. We use this range of numbers so the result of adding their number to 1969 is always between 2,000 and 2,999 which ensures the number in the thousands place is always a 2. This way when you strike it out and add it back in you always end up with 1998 which makes the answer 29 no matter what number they choose.

MATH FUNNIES

Dog Math:
One Pitbull asks another, "If I have 3 bones and Mr. Jones takes away 2, how many fingers will he have left?"

The Tower of Hanoi Trick

The Tower of Hanoi is a recreational mathematical puzzle popularized by the French mathematician Edouard Lucas in the year 1883. It consists of three pegs and many disks of different sizes you place in order of size on the pegs. The object of the puzzle is to move disks from the first peg arranged with the largest disk on the bottom and the smallest disk on top to the third peg without placing a larger disk on a smaller disk. You may remember this as a child's toy but I don't know too many children who can solve this puzzle, or at least one with more than three disks. I bought one from Amazon in the Toddler's Toys section with seven disks and if there is a child who can solve this puzzle they deserve a segment on the Ellen show with all expenses paid for their parents.

A Tower of Hanoi with 8 disks

If you want to predict how many moves it takes to move the disks from Peg A to Peg C there is a relatively simple solution. If there are n disks on Peg A the first operation is to transfer the top n – 1 disks from Peg A to B leaving C vacant; assume this takes x transfers. Next move the disk from the bottom of Peg A to C, this is why we say n – 1. Then reversing this process transfer the n – 1 disks from B to C, which will take x transfers. If it takes x transfers of disks to move a tower of n – 1 disks it will take 2x + 1 separate transfers of single disks to move a tower of n discs. This is what is called a "recursive pattern", it works but you need to know the number of moves of all the discs up to the one you are trying to solve:

n	x	2x + 1 = n
1	1	
2	3	2(1) + 1
3	7	2(3) + 1
4	15	2(7) + 1
5	31	2(15) + 1

Cumbersome! We need a way to solve for x by using n, or an *explicit* solution, not a *recursive* solution. So let's look at our results again and find a pattern.

n	x	
1	1	2
2	3	4
3	7	8
4	15	16
5	31	

The difference between each solution appears to be a multiple of 2! Can we find a relation between the powers of 2 and our values for n? When we substitute n for the power of 2 we see we're always off by 1.

$2^1 =$	2
$2^2 =$	4
$2^3 =$	8
$2^4 =$	16
$2^5 =$	32

If we subtract 1 from each of these answers we get the value for *x*, so let's try 2n - 1 as our explicit solution.

n	$2^n - 1$
1	1
2	3
3	7
4	15
5	31

It works! The explicit solution is $2^n - 1$, so the solution for any number of *n* disks is $2^n - 1$.

So my Tower of Hanoi with 7 disks would take me $2^7 - 1 = 127$ moves and an 8 disk version would take 255 moves.

MATH FUNNIES

I put my root beer in a square cup.

Now it's just beer.

A Mindreading Trick

For this trick I am going to read your mind!

- Think of a number between 1 and 10.
- Multiply it by 9 and add the digits of the answer together (if it's only 1 digit assume you're adding a zero).
- Subtract 6 from the number.
- Find the letter corresponding to the answer, A would be a 1, B is 2, etc.
- Think of a country whose name starts with the letter. I only want you to think of a country with a one-word name.
- Then think of a fruit whose name starts with the last letter of the country. This fruit should also only have a one-word name.

Go to the next page and prepare to have your mind blown!

Is the color of your fruit red? Is it an apple?

Whoa! How did I do that?

When you take a number between 1 and 10 and multiply it by 9 you always get a multiple of 9. When you add those digits together you're finding the digit sum which for multiples of 9 is always 9 (e.g., 9 x 4 = 36, 3 + 6 = 9). When you subtract 6 from your answer you will always get 3 which would correspond to the letter C.

Most countries starting with a C people will think of also end with an A: Canada, China, Cuba, Cambodia, and Croatia, for example. When asked to think of a fruit starting with the last letter of the country people will also most likely choose an apple, so if red isn't the color of their fruit ask them if they're thinking of a green apple or a Golden Delicious apple!

The only other common fruits starting with the letter A are apricots and avocadoes (yes, they're both fruits because they have seeds). If you didn't guess their fruit on the first try make like a psychic and say "Wait a minute, it's coming to me. I'm getting an image of an apricot. Is your fruit an apricot?"

If their fruit isn't an apricot just say "Somebody is thinking of apricots and it's throwing me off. Now I'm sensing a dark fruit, like an avocado. Is your fruit an avocado?"

If this doesn't work remind them you're only looking for countries and fruits with one-word names, so if they get cute and think of Congo as their country or açaí berries tell them those have two-word names and they weren't following instructions (Congo can refer to the Republic of the Congo or the Democratic Republic of the Congo).

If all this fails then congratulate them on having a truly difficult mind to read then try it again with someone else, I would say 9 times out of 10 this trick will work.

MATH FUNNIES

Why did the chicken cross the Mobius strip?

The Birth Month & Year Trick

Ask someone to multiply their birth month by 10, add 31 for the days in a long month, multiply the result by 10 again (just tell them to add a zero to their number), add the year they were born in two digits and subtract the 365 days in a year. With this information you can now tell them the month and year they were born! How does this work?

When you asked them to multiply their birth month by 10, add 31, multiply by 10 again, add their birth year and subtract 365 we could represent this as:

$$10m + 31$$
$$10(10m + 31)$$
$$10(10m + 31) + y - 365 = x$$
$$100m + 310 + y - 365 = x$$
$$100m + y - 55 = x$$

If you mentally add 55 to their result you get $100m + y$, making the digits on the left their birth month and digits on the right their birth year. Let's say they were born in July of 2004:

$$7 \times 10 + 31 = 101$$
add a 0 to get 1,010
add 04 and you get 1,014
subtracting 365 gives you 649

If you add 55 to 649 you get 704, and 7 is their birth of July and '04 is the year of their birth.

Let's try one more. Say a very awesome person was born in October of 1969. He would multiply the number of his birth month, 10, by 10 to get 100, add 31 to get 131, multiply by 10 again to get 1,310, add 69 to get 1,379, and then subtract 365 to get 1,014. Now all you have to do is add 55 to 1,014 to get 1,069 which tells us this their birth month was October and they were born in 1969.

There are many different ways you could go about doing this very old math trick. If you want to create your own twist on this trick like I did just come up with a different algebra expression. For example, if you didn't want to have to add 55 in your head you could choose an easier number by having them add 30 instead of 31 and subtract 360 degrees in a circle which would give you 60 to add to their number:

$$10(10m + 30) + y - 360 = x$$
$$100m + 300 + y - 360 = x$$
$$100m + y - 60 = x$$

The Lucky Numbers Trick

Pick any three-digit number and repeat the digits to get a six-digit number. Divide this number by 7, then 11, and finally 13 and you get your original number. Let's say our number is 357:

$$357{,}357 \div 7 = 51{,}051$$
$$51{,}051 \div 11 = 4{,}641$$
$$4{,}641 \div 13 = 357$$

How does this work for any three-digit number? The answer is repeating any three-digit number is the same as multiplying by 1,001, which is divisible by 7, 11, and 13. They are the prime divisors, or prime factors, quantities that will exactly divide the given quantity.

POP QUIZ FOR EXTRA CREDIT

Which is more correct to say,
5 and 8 <u>are</u> 14 or
5 and 8 <u>is</u> 14?

A Squaring Trick

Take any number and split it in two and the sum of those numbers' squares plus their individual products is always equal to the square of the original number:

```
     8 = 2 + 6                    12 = 8 + 4
  2²  =       4                 8²  =       64
  6²  =      36                 4²  =       16
  2x6 =      12                 8x4 =       32
  6x2 = +    12                 4x8 =  +    32
             64                            144
           √64  =  8                     √144  =  12


    24 = 18 + 6                   38 = 22 + 16
  18² =      324                22² =      484
   6² =       36                16² =      256
  18x6 =     108                16x22 =    352
  6x18 = +   108                22x16 =  + 352
            576                           1444
          √576  =  24                   √1444  =  38
```

MATH FUNNIES

Why was 6 afraid of 7?

The Sequential Remainders Trick

What number is it that when divided by 9 there shall remain 8, divide by 8 there shall remain 7, etc.?

This is another math riddle I found in a very old book dating back to the 1600's but it didn't provide a solution. I solved it by working the problem backwards and working my way up to a solution. Make a column of numbers 1 through 10, and starting with 0 in the top of the right column we need to multiply diagonally and then add up. We know when we divide 3 by 2 there will be a remainder of 1, 11 divided by 3 is 3 remainder 2, 47 divided by 4 is 11 remainder 3, etc. If we keep working our way up in this manner we will eventually get to the number that when divided by 9 there shall remain 8:

1		2 + 1 = 3
2	3	3 x 3 = 9, 9 + 2 = 11
3	11	11 x 4 = 44, 44 + 3 = 47
4	47	47 x 5 = 235, 235 + 4 = 239
5	239	239 x 6 = 1,434, 1,434 + 5 = 1,439
6	1,439	1,439 x 7 = 10,073, 10,073 + 6 = 10,079
7	10,079	10,079 x 8 = 80,632, 80,632 + 7 = 80,639
8	80,639	80,639 x 9 = 725,751, 725,7561 + 8 = 725,759
9	725,759	725,759 x 10 = 7,257,590, 7,257,590 + 9 = 7,257,599
10	7,257,599	

Now we have a number that when divided by 9 there shall remain 8, and divide by 8 there shall remain 7, and so on.

```
       80,639              10,079              1,439
   9)725,759            8)80,639            7)10,079
      72                   8                   7
      05                   063                 30
       0                    56                 28
       57                   79                 27
       54                   72                 21
       35                    7                 69
       27             remainder = 7!           63
       89                                       6
       81                                 remainder = 6!
        8
 remainder = 8!
```

If we keep working our way up the list the remainder of each division will be one less than the previous remainder.

Numerical Tricks

Numbers can do amazing things when you combine them in the right way. Take these examples, for instance:

$$1 \times 8 + 1 = 9$$
$$12 \times 8 + 2 = 98$$
$$123 \times 8 + 3 = 987$$
$$1234 \times 8 + 4 = 9876$$
$$12345 \times 8 + 5 = 98765$$
$$123456 \times 8 + 6 = 987654$$
$$1234567 \times 8 + 7 = 9876543$$
$$12345678 \times 8 + 8 = 98765432$$
$$123456789 \times 8 + 9 = 987654321$$

$$9 \times 9 + 7 = 88$$
$$98 \times 9 + 6 = 888$$
$$987 \times 9 + 5 = 8888$$
$$9876 \times 9 + 4 = 88888$$
$$98765 \times 9 + 3 = 888888$$
$$987654 \times 9 + 2 = 8888888$$
$$9876543 \times 9 + 1 = 88888888$$
$$98765432 \times 9 + 0 = 888888888$$
$$987654321 \times 9 - 1 = 8888888888$$
$$9876543210 \times 9 - 2 = 88888888888$$

$$1 \times 9 + 2 = 11$$
$$12 \times 9 + 3 = 111$$
$$123 \times 9 + 4 = 1111$$
$$1234 \times 9 + 5 = 11111$$
$$12345 \times 9 + 6 = 111111$$
$$123456 \times 9 + 7 = 1111111$$
$$1234567 \times 9 + 8 = 11111111$$
$$12345678 \times 9 + 9 = 111111111$$
$$123456789 \times 9 + 10 = 1111111111$$

$$11^2 = 121$$
$$111^2 = 12321$$
$$1111^2 = 1234321$$
$$11111^2 = 123454321$$
$$111111^2 = 12345654321$$
$$1111111^2 = 1234567654321$$
$$11111111^2 = 123456787654321$$
$$111111111^2 = 12345678987654321$$

$$1 + 2 + 1 = 2^2$$
$$1 + 2 + 3 + 2 + 1 = 3^2$$
$$1 + 2 + 3 + 4 + 3 + 2 + 1 = 4^2$$
$$1 + 2 + 3 + 4 + 5 + 4 + 3 + 2 + 1 = 5^2$$
$$1 + 2 + 3 + 4 + 5 + 6 + 5 + 4 + 3 + 2 + 1 = 6^2$$
$$1 + 2 + 3 + 4 + 5 + 6 + 7 + 6 + 5 + 4 + 3 + 2 + 1 = 7^2$$
$$1 + 2 + 3 + 4 + 5 + 6 + 7 + 8 + 7 + 6 + 5 + 4 + 3 + 2 + 1 = 8^2$$
$$1 + 2 + 3 + 4 + 5 + 6 + 7 + 8 + 9 + 8 + 7 + 6 + 5 + 4 + 3 + 2 + 1 = 9^2$$

$$121 = \frac{22 \times 22}{1+2+1}$$

$$12321 = \frac{333 \times 333}{1+2+3+2+1}$$

$$1234321 = \frac{4444 \times 4444}{1+2+3+4+3+2+1}$$

$$123454321 = \frac{55555 \times 55555}{1+2+3+4+5+4+3+2+1}$$

$$12345654321 = \frac{666666 \times 666666}{1+2+3+4+5+6+5+4+3+2+1}$$

$$1234567654321 = \frac{7777777 \times 7777777}{1+2+3+4+5+6+7+6+5+4+3+2+1}$$

$$1234567654321 = \frac{7777777 \times 7777777}{1+2+3+4+5+6+7+6+5+4+3+2+1}$$

$$123456787654321 = \frac{88888888 \times 88888888}{1+2+3+4+5+6+7+8+7+6+5+4+3+2+1}$$

$$12345678987654321 = \frac{999999999 \times 999999999}{1+2+3+4+5+6+7+8+9+8+7+6+5+4+3+2+1}$$

MATH FUNNIES

Dear Algebra,

Please stop asking us to find your X.

She's never coming back, and we don't know Y.

The Number 9 Trick

In Chinese and other Asian cultures the number 8 is considered a magic number but this trick demonstrates how the number 9 may be even more magical.

Take the number series 12345679 (1 through 9 without the number 8) and multiply it by any multiple of 9:

$$
\begin{array}{r}
12{,}345{,}679 \\
\times 9 \\ \hline
111{,}111{,}111
\end{array}
$$

$$
\begin{array}{r}
12{,}345{,}679 \\
\times 18 \\ \hline
222{,}222{,}222
\end{array}
$$

$$
\begin{array}{r}
12{,}345{,}679 \\
\times 27 \\ \hline
333{,}333{,}333
\end{array}
$$

$$
\begin{array}{r}
12{,}345{,}679 \\
\times 36 \\ \hline
444{,}444{,}444
\end{array}
$$

You can do this with all the multiples of 9 up to 81. After that the pattern changes:

$$
\begin{array}{r}
12{,}345{,}679 \\
\times 90 \\ \hline
1{,}111{,}111{,}110
\end{array}
$$

Here are a few more number 9 tricks:

$$\begin{array}{r} 370{,}370{,}37 \\ \times \phantom{000{,}00}9 \\ \hline 333{,}333{,}333 \end{array}$$

$$\begin{array}{r} 370{,}370{,}37 \\ \times \phantom{000{,}00}18 \\ \hline 666{,}666{,}666 \end{array}$$

$$\begin{array}{r} 370{,}370{,}37 \\ \times \phantom{000{,}00}27 \\ \hline 999{,}999{,}999 \end{array}$$

$$\begin{array}{r} 1{,}371{,}742 \\ \times \phantom{000{,}00}9 \\ \hline 12{,}345{,}678 \end{array}$$

$$\begin{array}{r} 98{,}765{,}432 \\ \times \phantom{000{,}00}9 \\ \hline 888{,}888{,}888 \end{array}$$

MATH FUNNIES

What do you call friends who love math?

The Card Whisperer Trick

With this card trick you will allow someone to rearrange and remove stacks of cards but no matter how hard they try you will always be able to predict the remaining top card, after consulting with the other cards of course.

Shuffle a deck of 52 cards and turn over the top card. Starting with the number of the card count to 14 and add a card each time. For example, if the top card was a 6 you would add 8 cards to the pile. Aces count as 1, Jacks are 11, Queens are 12 and Kings are 13. Do this three more times until you have four piles of cards. They can then rearrange the stacks of cards as much as they want so you won't be able to tell which stack was which. Have them remove two of the stacks and add them to the other unused cards we'll call the discard pile. Now tell them if they flip over either of the top cards you will predict the value of the face-down top card after you consult with the cards in the discard pile. You then touch all the cards in the discard pile to read their psychic energy and much to your audience's surprise you successfully predict the value of the remaining top card. Are you really psychic? How did you know the value of the card?

The answer is we use a little algebra. Since we count to 14 we can say each of the piles we'll call a and b are equal to 15 minus some number:

$$a = 15 - x$$
$$b = 15 - y$$

We also know the number of cards in the discard pile are equal to 52 minus the number of cards in the two piles. We'll call the discard pile c:

$$c = 52 - a - b$$
$$c = 52 - (15 - x) - (15 - y)$$
$$c = 52 - 15 + x - 15 + y$$
$$c = 22 + x + y$$
$$x = c - 22 - y$$
$$y = c - 22 - x$$

As we're touching each of the discard pile cards we're really counting them. The value of the face-down top card is the number of discard pile cards less 22 and the value of the face-up top card. So, if the face-up card was the 4 of Clubs you would silently count out twenty-two cards from the discard pile, then count four more, and if you only had one card left you would announce "The cards have revealed to me the other top card is an Ace!"

A variation of this trick would be to flip over the top card of a well-shuffled deck and add cards until you get to thirteen. Flip the stack over and repeat until you can't make any more piles. Ask someone to pick three of the stacks and put the other cards in the discard pile. Then have them flip over the top card of two of the decks and tell them the cards will tell you the value of the face-down top card. This time though you will count ten cards in the discard pile, then a number of cards equal to the face values of the two face-up cards, and the remaining number of cards will be the value of the face-down card. If we call the stacks *a*, *b*, and *c* we can say their values are 14 minus some number:

$$a = 14 - x$$
$$b = 14 - y$$
$$c = 14 - z$$

We can then say the value of the discard pile we'll call *d* is equal to 52 less the values of *a*, *b*, and *c*:

$$d = 52 - a - b - c$$
$$d = 52 - (14 - x) - (14 - y) - (14 - z)$$
$$d = 52 - 14 + x - 14 + y - 14 + z$$
$$d = 10 + x + y + z$$

When they flip over *y* and *z* you know *x* is

$$x = d - 10 - y - z$$

Then it's just a matter of counting out ten cards from the discard pile and the same number of cards equaling the values of y and z to get the value of the face-down top card, x.

Here's a little advice if you plan on doing these kind of card tricks. Math-based card tricks, or "self-working" card tricks, can be some of the best card tricks in the world but also some of the most tedious and boring. A little playful banter goes a long way to keep your audience from getting bored or distracted. Practice until you are confident enough to do the trick and talk at the same time. While you are secretly counting the cards pause occasionally and go "hmmm, interesting" or "Oh, really?" as if you are communicating with the cards.

MATH FUNNIES

What do you call a number that can't sit still?

The Gergonne Card Trick

This trick is credited to the French mathematician Joseph Diaz Gergonne (pronounced "Yo-sef Dee-az Jair-gone). He was an officer in the French army during the French Revolution before devoting his life to mathematics. When he couldn't get his work published in traditional publications he started his own mathematics journal called *Annales de mathématiques pures et appliquées*, also simply known as *Annales de Gergonne*, where he published 200 of his own articles and those of the most distinguished mathematicians of his time. It was in one of these journals he published a detailed analysis of this trick.

Ask someone to pick a number between 1 and 27 and tell you their number, or you can ask their age and use it instead (if they're under the age of 28). Deal out three piles of nine cards face up and ask them to pick one of those cards but not tell you what it is, just point at the stack where the card is located. Repeat two more times with them pointing at the stack containing their card each time, and after the third time the position of their card in the pile will be the same as their number! How did Monsieur Gergonne come up with this trick?

The secret is to stack their pile in an order based on their chosen number using the ternary numeral system, base 3. This sounds terribly complicated but all you need to do is learn this table:

3^0	3^1	3^2	
0	0	0	Top
1	3	9	Middle
2	6	18	Bottom
1st	2nd	3rd	

Any number raised to the zero power is 1, and any number raised to the first power is the number. This is how we get the middle numbers, 3 to the zero power is 1, 3 to the first power is 3, and 3 squared is 9. The top row is always 0 and the bottom row is the middle number multiplied by 2. If we're showing this trick to a 12-year-old and using their age for our number we would subtract 1 from 12 to get the number of cards sitting on top of their card the last time we stack them. Then we use the table to see what combination of numbers will give us 11. The largest number before 11 in the table is 9 which we see is in the 3rd row. To make 11 we also need a 2 which we see is in the first row. Going from left to right we need a 2, 0, and 9 to make 11 and we see 2 is on the bottom, 0 is on top, and 9 is in the middle. This is the order of where to place their stack each time, first put it on the bottom, next put it on top, and the last time place it in the middle. When you flip the cards over for the last time you can then casually say "How old were you again? 12?" and the 12th card you count out will magically be their number.

A variation of this trick would be to place their card in the middle each time, and their card will always be the middle card after the third time. Since we're not using base 3 we don't have to use 27 cards (3^3), this is sometimes called the 21 Card Trick but you can use as few as twelve cards and still make it work. Just figure out what the middle card is based on the number of cards you're using, if you're using twelve cards their card will be the sixth in the deck, if you're using fifteen their card will be the eighth, etc.

MATH FUNNIES

If you have 7 apples in one hand and 5 apples in the other what do you have?

Howard Adams Mated Cards Trick

This is my version of a card trick created by magician Howard Adams (1931-2010) and published in his book *OICUFESP* (Oh I See You Have ESP). Randomly select six different cards and find their mates, cards of the same number or face and color, like the 3 of Hearts and the 3 of Diamonds. Arrange one set of cards vertically and face-up and place their mate next to them face-up. Pick up one column of cards starting with the bottom card and work your way up, do the same with the other column, and place the second stack of cards under the first. You or a spectator can then cut the twelve cards as many times as you like. Split the stack in half and flip the right stack of cards face up. Both stacks will then be randomly shuffled using the magic phrase "CARDS WILL MIX BY MAGIC", and after each shuffle take the top card from each stack and set them aside without flipping them over. For example, you could pick up the face down deck, take the card off the top of the deck and move it to the bottom of the deck for the letter "C". Repeat two more times for the letters "A" and "R". Then pick up the face-up stack and move the top card to the bottom for the letter "D" and do the same for the letter "S". Or you could use just the face-down stack or just the face-up stack, or alternate stacks each time, it doesn't matter. When you're done with each word take the top card from the face down deck and set it aside and put the top card from the face-up deck next to it. Repeat this process for each of the words "WILL", "MIX", "BY", and "MAGIC", each time placing the top cards aside and next to each other. After you finish shuffling with all the "magic words" you should have two columns of six cards with the left column face-up and the right column face-down.

You can now tell your spectators something magical has happened! Even though you cut the deck a random number of times and randomly selected cards to shuffle with the "magic words" the mated cards have mysteriously found their way back to each other. As you flip over each face down card your audience will be amazed to see the cards are with their mates again. But how?

It isn't the magic words but rather a predetermined way of rearranging the cards a certain number of times disguised as words. "CARDS" is 5 because it has five letters, "WILL" is 4, etc. The numbers come from the formula $x \text{ MOD } n = n - 1$ where x is the number of letters in each word and n is the number of cards in each of the two stacks. MOD is short for Modulo Operation which is used in computer science to find the remainder after the division of one number by another (called the "modulus"). Using two positive numbers a (the dividend) and n (the divisor) we would write this as a MOD $n = b$. When we had six cards this could be expressed as 5 MOD 6 = 5 and 11 MOD 6 = 5, because 5 divided by 6 is 0 with a remainder of 5, and 11 divided by 6 is 1 with a remainder of 5.

For the first shuffle using the word "CARDS" there are five letters and six cards in each stack, so substituting 5 for x and 6 for n we get 5 MOD 6 = 5. Using this formula we can come up with other words and phrases:

	Cards in the Pile	Word Length
8	x MOD 8 = 7	x = 7
7	x MOD 7 = 6	x = 6, 13
6	x MOD 6 = 5	x = 5, 11
5	x MOD 5 = 4	x = 4, 9, 14
4	x MOD 4 = 3	x = 3, 7, 11
3	x MOD 3 = 2	x = 2, 5, 8
2	x MOD 2 = 1	x = 1, 3, 5

Now you can see why we chose the word "CARDS" with 5 letters when we had 6 cards, the word "WILL" with 4 letters when there were 5 cards, etc. For eight cards we could have used a word with a length of 23 letters but there are only thirteen English words with 23 letters and all of them are ridiculous (e.g., "indistinguishablenesses"). If we wanted to take this to the next level we could use 8 cards and the phrase "AGAINST BELIEF CARDS WILL MIX BY MAGIC", and if you look at the table you will see why I chose these words. However, this would place the trick squarely in bore-your-audience-to-death territory so I chose to stop at 6 words, and maybe Mr. Adams was right to only use 5 words. I'll let you decide.

If you made it this far you might be thinking to yourself, "This is terribly fascinating but what does it have to do with the cards matching back up?" As with most card and magic tricks we use deception and misdirection. When you or your spectator repeatedly cut the deck all you were doing is cycling the cards, not mixing them up. Splitting the deck in half left each stack with one of the mated pairs, and they were still in order until you flipped the right stack face-up. If we started with black Aces, Kings, Queens, Jacks, Tens, and Nines, when we put the cards together they would be in the order of 9, 10, J, Q, K, A, 9, 10, J, Q, K, A, from top to bottom. When we made cuts to the deck we rearranged the order but the mates are still equally spaced, for example the order may now be K, A, 9, 10, J, Q, K, A, 9, 10, J, Q. Splitting the deck in half makes the left stack K, A, 9, 10, J, Q and the right stack K, A, 9, 10, J, Q, and turning the right stack face-up reverses its order. In this scenario, we have a black King facing up on the right and a black King on the bottom of the left stack. If we were to move the top card on the left to the bottom five times the King on the bottom of the left stack would move to the top of the stack, exactly where the face-up King in the right stack is located. If we had chosen to move the top two cards on the left to the bottom of the stack and the top three cards on the right to the bottom we would now have the 10's mated. Any combination of moving cards from the top of the stack to the bottom will result in a mated pair as long as we only move them the right number of times we calculated with our Modulo Operation.

The beautiful thing about this card trick is if you understood nothing I just wrote about how this trick works you can *still* do it and amaze people with your mathematical abilities. The beauty of self-working card tricks is they work every time without having to resort to any trickery. If you do understand the math this is also a great way to get someone interested in mathematics, and it's much more entertaining than watching someone talk to a whiteboard while they write down equations.

A Classroom Card Trick

Walk around the classroom and have students pick one card each from the deck. Have them keep their cards secret and do the following math:

1. Write down the face value of the card. Aces are 1, Jacks are 11, Queens are 12, and Kings are 13.
2. Double the face value.
3. Add 3.
4. Multiply by 5.
5. If the card you selected is a Club, add 1; a Diamond, add 2; a Heart, add 3; or a Spade, add 4 (to make this easier to remember they're in alphabetical order).
6. Ask each participant the result of their math and you will not only tell them their card but also their suit!

You determine their card and suit by subtracting 15 from their result, the number on the left gives you the face value of the card and the number on the right gives you the suit. We can represent what we did algebraically:

- c equals the face value of the card
- $2c$ is double the face value
- adding 3 gives us $2c + 3$

- multiplying by 5 is 10c + 15
- s = suit so 10c + 15 + s
- subtracting 15 leaves 10c + s

If one of the cards was a King of Spades you would have:

- 13 x 2 = 26
- 26 + 3 = 29
- 29 x 5 = 145
- 145 + 4 = 149
- 149 – 15 = 134
- 13 tells us the card is a King and 4 tells us it's a Spade

This is a simple enough trick you could give a class of 25 kids two cards each and "guess" each card correctly (just don't forget to take out the Jokers!). When they ask you the secret to the trick they'll be surprised to learn the trick is actually math, which you tricked them into learning!

MATH FUNNIES

Bob has 38 candy bars and eats 30. What does he have now?

The Magic Age Table Trick

There is a very old math trick where you ask someone to identify the columns in a table where their age can be found and with just this information you can tell them their age.

Let's say they tell you their age can be found in columns 2, 4, and 6 of the table on the next page. Much to their amazement you immediately tell them their age is 42! How did you know? Did you memorize the entire table? You could tell them you memorized the table but the real trick is to add the numbers from the first row in the columns they told you contained their age. In this case those numbers are 2, 8, and 32, and 2 + 8 + 32 = 42.

The thing is in all the references I've found to this old trick no one has ever explained why it works. So lucky for you I've reverse engineered the Magic Age Table trick and have an explanation for you on the page after the Table. Copies of the Magic Age Table and the solution can be downloaded here:

https://KidsSci.com/CrushYourMathFear/MagicAgeTable.pdf

https://KidsSci.com/CrushYourMathFear/MagicAgeTable2.pdf

Magic Age Table

1	2	3	4	5	6
1	2	4	8	16	32
3	3	5	9	17	33
5	6	6	10	18	34
7	7	7	11	19	35
9	10	12	12	20	36
11	11	13	13	21	37
13	14	14	14	22	38
15	15	15	15	23	39
17	18	20	24	24	40
19	19	21	25	25	41
21	22	22	26	26	42
23	23	23	27	27	43
25	26	28	28	28	44
27	27	29	29	29	45
29	30	30	30	30	46
31	31	31	31	31	47
33	34	36	40	48	48
35	35	37	41	49	49
37	38	38	42	50	50
39	39	39	43	51	51
41	42	44	44	52	52
43	43	45	45	53	53
45	46	46	46	54	54
47	47	47	47	55	55
49	50	52	56	56	56
51	51	53	57	57	57
53	54	54	58	58	58
55	55	55	59	59	59
57	58	60	60	60	60
59	59	61	61	61	61
61	62	62	62	62	62
63	63	63	63	63	63

Magic Age Table

Notice the 1st column skips every other number up to the number 63.

The 2nd column starts with the first number skipped in the 1st column, the number 2. The 2nd column then skips every other 2 numbers up to 63.

The 3rd column starts with the 1st number skipped in the 2nd column, 4, and then skips every other 4 numbers up to the number 63.

The 4th column starts with the 1st number skipped in the 3rd column, 8, then skips every other 8 numbers up to the number 63.

Are you seeing a pattern yet?

By arranging the columns in this manner you can represent any age up to 63 with combinations of 1, 2, 4, 8, 16, & 32 (powers of 2 from 2^0 to 2^5).

1	2	3	4	5	6
1	2	4	8	16	32
3	3	5	9	17	33
5	6	6	10	18	34
7	7	7	11	19	35
9	10	12	12	20	36
11	11	13	13	21	37
13	14	14	14	22	38
15	15	15	15	23	39
17	18	20	24	24	40
19	19	21	25	25	41
21	22	22	26	26	42
23	23	23	27	27	43
25	26	28	28	28	44
27	27	29	29	29	45
29	30	30	30	30	46
31	31	31	31	31	47
33	34	36	40	48	48
35	35	37	41	49	49
37	38	38	42	50	50
39	39	39	43	51	51
41	42	44	44	52	52
43	43	45	45	53	53
45	46	46	46	54	54
47	47	47	47	55	55
49	50	52	56	56	56
51	51	53	57	57	57
53	54	54	58	58	58
55	55	55	59	59	59
57	58	60	60	60	60
59	59	61	61	61	61
61	62	62	62	62	62
63	63	63	63	63	63

By using this pattern you can find ages up to 63, if you added another column you could go up to 127 years old!

Math Riddles

Many of you are going to flip to this section and say, "This book is full of math problems!" I choose to call them riddles because they need to be solved, and calling them problems has a negative connotation. This is no different than when people say, "There are no problems, only opportunities". There are 100 math riddles in all and if you can master them you will be able to solve most algebra and trigonometry based story questions. Even the dreaded "If Train A leaves the station at…" type questions. If you get stumped there are detailed explanations for how to solve every riddle. For some you only need an elementary level understanding of mathematics and others require a basic knowledge of algebra, geometry, and trigonometry. There are even a few logic problems for you to try.

1. The difference between two numbers is 40; the difference in their squares is 4,800. What are the numbers?

2. A number increased by its cube is 592,788, what is the number?

3. The cube root of a certain number is 10 times the 4th root. What is the number?

4. If 2 miles of fence will enclose a square of 160 acres, how large a square will 3 miles of fence enclose?

5. You buy $90 of stock in a company, sell your shares for $100, and repurchase shares for $80. How much did you make by trading?

6. If the Earth is 7,918 miles in diameter and the Sun is 865,000 miles in diameter how many Earths would it take to equal the Sun?

7. The sum of two numbers is 582 and their difference is 218. What are the numbers?

8. A dozen writing pads and a box of pens cost $17.18. If the writing pads are $6.20 more than the pens how much do both items each cost?

9. Michael said to Andrew "Give me $10.00 and I will have as much money as you." Andrew said to Michael "Give me $10.00 and I will have twice as much money as you." How much money do Michael and Andrew have each?

10. The product of two numbers exceeds their difference by their sum. Find one of these numbers.

11. Twice the sum of two numbers plus twice their difference is 80. What is the greater number?

12. One half the sum of two numbers exceeds one half their difference by 60. What is the smaller number?

13. When Sally married Harry she was $5/6$ his age and 24 years later she was $11/12$ his age. How old were Harry and Sally when they married?

14. Sum to infinity the series 1 + ½ + ¼ + ⅛ ...

15. Find the sum of the series 4 + .4 + .04 + .004 + ... and express it as a fraction.

16. Find the volume of a rectangular piece of ice 8 feet long, 7 feet wide, and floating in water with 2.4 inches of its thickness above the water, assuming the density of ice is 0.9 (as compared to water).

17. 15,600 is the product of three consecutive numbers. What are they?

18. Find a number which is as much greater than 1,042 as it is less than 1,236.

19. What is the smallest number to be subtracted from 10,697 to make a perfect cube?

20. A karate class instructor has each of his students bow to every other student in the class and to him. In one of his classes where there are twice as many girls as boys 900

bows were made. How many boys are in the class?

21. A boy asks a mathy girl her age and she responds, "If you take the square root of my age and add it to ⅜ of my age you'll get 10." How old is she?

22. There is a fish whose head is 9 inches long, the tail is as long as the head plus one half the body, and the body is as long as the head and tail together. How long is the fish?

23. Mary is 24 years old. She is twice as old as Ann was when Mary was as old as Ann is now. How old is Ann?

24. If 6 cats can eat 6 rats in 6 minutes, how many cats will it take to eat 100 rats in 100 minutes?

25. A boy was sent to get 4 quarts of water for a science experiment but all he had was a 5-quart and 3-quart container. How does he get exactly 4 quarts?

26. A frog falls into a well 45 feet deep. Every day he climbs 3 feet up but slides 2 feet back. How many days does it take

him to escape the well?

27. A man and his wife and two sons must cross a river in a boat only capable of carrying 150 pounds. If the man and wife both weigh 150 pounds and the sons each weigh 75 pounds how do they all get across?

28. An aquarium filled to the brim weighs 20 pounds. If we put a live 5-pound fish in the tank does its weight increase or decrease?

29. A man lived ¼ his life as a boy, ⅕ as a youth, ⅓ as a young man, and 13 years as a retiree. How old is he?

30. Andrew, Brandon, and Connor are going on a school trip. Andrew is bringing 20% more money than Brandon, Brandon is bringing 25% more than Connor, and Andrew will have $80 more than Connor. How much money are the boys taking on the school trip?

31. A grandfather living with his son, his son's wife, and their three sons lived to 125 years old, which coincidentally happened to be his family's combined ages. The father

said he was as old as his wife and second oldest son plus one year, the wife said she was as old as all three of her sons plus three years, the oldest son said he was as old as both his younger brothers plus three years, and the second oldest son said he was four times as old as his younger brother plus one year. How old is each family member?

32. A Texas farmer keeps 2,100 cows on his farm. For every 3 cows he plows 1 acre of ground and for every 7 cows he pastures 2 acres of land. How many acres is his farm?

33. When gold was worth 25% more than paper money what was the value in gold of a dollar bill?

34. Cody is riding his bicycle cross country and plans to go a total of 1,120 miles. The first day he will cycle for 40 miles and then add a set number of miles each day until he reaches 100 miles per day, at which time he will have reached his destination. What color are his socks? Just kidding, what we want to know is how many days will he by cycling?

35. A 120-foot-tall tree was broken in a storm and the top of the tree struck the ground 40 feet from the base of the tree. How long was the fallen part of the tree?

36. If a log starts from the source of a river on Friday and floats 80 miles downstream during the day but comes back 40 miles during the night with the return tide, on what day of the week will it reach the mouth of the 300-mile-long river?

37. If a ball of yarn 6 inches in diameter makes one pair of gloves how many similar pairs can you make with a 12-inch diameter ball of yarn?

38. Find the sum of 2,324 thousandths and 24,325 hundredths.

39. The taekwondodos are fighting over the value of two melons. If one melon with a diameter of 20 inches sells for $2.00 what is the cost of a 30-inch diameter melon?

40. From 200 hundredths take 15 tenths.

41. A crew can row 24 miles downstream in 3 hours but requires 4 hours to row back. What is the rate of the current?

42. What minuend is 80 more than the subtrahend which is 20 more than the remainder?

43. Logan makes a wager with Jeff Bezos for every free throw he makes he will win a penny for the first shot, 2 cents for the second shot, 4 cents for the third shot, and so on, doubling the winnings each time. Logan makes 32 free throws so how much does Jeff Bezos owe him?

44. If sound travels at the rate of 1,125.3 feet per second how far distant is a thunder cloud when the sound of thunder follows the flash of lightning by 10 seconds?

45. How much will the product of two numbers be increased by increasing each number by 1?

46. What is the total distance a ball travels before coming to rest if it is dropped from a height of 100 feet and bounces

half the distance after each fall?

47. Sean needs to be at school by a certain time. If he walks 4 miles per hour he will be 10 minutes late. If he walks 5 miles per hour he will be 20 minutes early. How far does he have to walk?

48. A boat can go 20 miles an hour downstream but only 15 miles an hour upstream. If the boat takes 5 hours longer going upstream than downstream how far did it go?

49. Marliesa is 30 years old and Jasmine is 3 years old. In how many years will Marliesa be 5 times older than Jasmine?

50. David can row upstream in 3 hours and downstream in 2. If the rate of the current is 1 mile per hour how far does he row?

51. A quarterback is 27 steps ahead of the opposing teams nearest defensive player and he thinks he's going to make a touchdown since 8 of his steps is equal to 5 of the linebacker's steps. The only problem is the linebacker is a giant and every 2 steps he takes equals 5 of the

quarterback's steps. How many steps before the linebacker tackles the quarterback?

52. Sarah asked Morgan how many fish she caught and she replied, "11 fish are 7 more than $\frac{2}{5}$ the number of fish I caught." How many fish did Morgan catch?

53. Time for some mental arithmetic! Quickly sum the following numbers in your head:

$1{,}000 + 40 + 1{,}000 + 30 + 1{,}000 + 20 + 1{,}000 + 10 = ?$

54. I am now twice as old as you were when I was your age. The sum of our ages is 63. What are our ages?

55. A military officer wanted to arrange her soldiers in a square formation but on her first try found she had 39 soldiers left over. On her second try she increased the number of soldiers on each side by one and found she was short 50 soldiers. How many soldiers are there?

56. A 96-foot-tall tree was broken in a wind storm and the top of the tree fell over with the tip of the tree landing 36

feet from the base. If the two parts of the tree did not separate how tall is the stump?

57. You need a piece of wood 9 inches by 16 inches but all you have is a 12-inch square board. How do you make this piece of wood work for you?

58. Zach, Dylan, and Cole found $60 and decided the fairest way to split the money up was for Zach to get ⅓, Dylan to get ¼, and Cole to get ⅕. How much money did each receive?

59. A room with 8 corners had a cat in each corner, 7 cats before each cat, and a cat on every cat's tail. How many cats are in the room?

60. A man with a fox, a goose, and a bushel of corn needs to get across the river, but with his tiny boat he can only take them across one at a time. The fox will kill the goose and the goose will eat the corn if left together. How can he get them across safely?

61. You need to weigh many items ranging in weight from 1 to 40 pounds but all you have access to is a balance scale. To make matters worse you can only take 4 weights with you. With what 4 weights can you weigh any number of pounds from 1 to 40? Everything you'll be weighing is a whole number.

62. You have $10,000. If you spend half the money today and half the remainder each following day in how many days will you have no money?

63. With the nine digits 9, 8, 7, 6, 5, 4, 3, 2, and 1 express 4 numbers whose sum is 100 using each digit only once. Powers, roots and fractions are allowed.

64. Starting with a square of 12 pennies with 4 pennies on each side can you arrange the pennies so there are 5 pennies on each side without adding any pennies or changing the shape?

65. With the ten digits 9, 8, 7, 6, 5, 4, 3, 2, 1, and 0 write an expression with three numbers whose sum is 4 ½.

Powers, roots and fractions are allowed.

66. You were hired by a home renovation show on the House & Yard TV channel to make a 2-foot by 15-foot table. You ordered a 2-foot by 15-foot plank from your lumber supplier but they sent you a 3-foot by 10-foot plank by mistake. The big reveal is in an hour and the lumber yard can't get you the right sized piece of wood in time. You tell the show's host this horrible development and she tells you to just make one cut to the plank and it will be fine, then runs off without further explanation. What do you do?

67. Write an expression equaling 24 with three identical numbers, none of which are 8. There are 2 possible answers.

68. You are on a cruise ship heading due south traveling 30 knots (about 35 miles per hour). The ship is 850 feet long and you are standing 50 feet from the bow (the front of the ship). In just a few seconds you are going to cross the equator, at which time you are going to make a great leap. Which way will you jump further, in the direction of the

ship, or in the opposite direction (to the north), assuming you jump the same distance in both directions?

69. You hired a contractor to install a security system for your stables to keep track of everyone's comings and goings but discovered the system only counts heads and feet. This may be fine for people-only facilities but you need to track people and horses. You run your first report and see there are 82 feet and 26 heads in the stable area, how many horses and riders are there?

70. Arrange the numbers 1, 2, 3, 4, 5, 6, 7, 8, and 9 so their sum will be 100 and all the numbers are used only once. There are several possible solutions so use your imagination!

71. At 10 am a train leaves New York City for Washington DC traveling at 50 miles per hour. The conductor will be standing at the tail end of the train. At the same time, another train leaves Washington DC for New York City traveling at 40 miles per hour. Its conductor will be seated in the center of the train. When the two trains meet and the conductors wave at each other which one

will be closer to New York City?

72. Find all the missing numbers:

```
      1  _  3  2  2  7  1
               5  2  _  4
         6  3  _  _  7  4
            8  8  _  4  7
            3  0  5  4  1  7
   +  2  _  3  5  4  7  _
   ─────────────────────────
      4, 1  0  7, 3  0  3
```

73. Strike out nine of the numbers below so the total of the remaining numbers is 1,111:

```
       1  1  1
       3  3  3
       5  5  5
       7  7  7
    +  9  9  9
```

74. Huey, Dewey, and Louie were packing their lunches and ran into a problem: their uncle said they could split a 24-ounce bottle of juice but they only had a 5-ounce, 11-ounce, and 13-ounce bottle to take with them. How do

they equally divide the juice?

75. You plant a grain of corn 5 inches under the soil. The first night it sprouts and grows ½ the distance to the surface of the earth and continues to grow ½ the remaining distance each night after. How long before its shoots pop out of the soil?

76. David is 71 and his son Chris is 34. At what age was David three times as old as Chris?

77. At what age will Chris be half his father's age?

78. What two-digit number is equal to 5 times the sum of its digits, and when 9 is added to the number the digits of the number reverse?

79. Express the numbers 1 to 30 using only 4's. There are many solutions, here are some examples:

$$50 = (4 \times 4 + 4) / 0.4$$
$$46 = 44 + 4 / \sqrt{4}$$

80. Express the numbers from 1 to 30 using the first four prime numbers in order. Roots, powers, and factorials are acceptable. There are many solutions, for example:

$$46 = 2^3 + 3! + 5^2 + 7$$

81. A man had 17 cows. He died. His will stated his property should be divided as follows:
  - the first son gets ½ of the cows,
  - the middle son gets ⅓ of the cows,
  - and the youngest son gets ⅑ of the cows.

  At first the sons couldn't figure out how to divide the cows without turning them into hamburger. A neighbor volunteered to help and loaned the sons one of his cows, and the first son took home 9 cows, the middle son took home 6 cows, and the youngest son took home 2 cows. Then they returned the cow to their neighbor. How did they make this work?

82. You're working for the home renovation show on HYTV (House & Yard Television) again and are asked to make a custom end table. You ordered a 27-inch square slab of exotic bubinga wood from Central Africa through your

favorite lumber supplier but instead they sent you a 12-inch by 60-inch piece of bubinga wood. You tell the host of the show your big problem and she says no problem, just make 4 cuts and rearrange the pieces and it will be fine. How do you turn a 12-inch by 60-inch piece of wood into a 27-inch square?

83. Solve the following problem: if each letter corresponds to a number 0 to 9 what is:

$$\begin{array}{r} A\ B\ C \\ +\ D\ E\ F \\ \hline G\ H\ I\ J \end{array}$$

There are many solutions.

84. The previous problem is an example of an alphametic puzzle, also known as alphametics, cryptarithmetic, verbal arithmetic or word addition. This problem is probably the most famous alphametic puzzle, created by the famous puzzle maker H.E. Dudeney and published in *Strand Magazine* in July 1924:

$$\begin{array}{r} S\ E\ N\ D \\ +\ M\ O\ R\ E \\ \hline M\ O\ N\ E\ Y \end{array}$$

This alphametic has only one solution, puzzles like this are said to have unique solutions.

85. This is another famous alphametic puzzle created by Peter Macdonald and published in the *Journal of Recreational Math* in 1977, and like the previous alphametic it has only one solution:

$$\begin{array}{r} Z E R O E S \\ + O N E S \\ \hline B I N A R Y \end{array}$$

86. Alphametic puzzles whose letters have meaning separate from the solution of the puzzle are called double-unique solutions or doubly-true solutions. The addends and the sum are "number words", and when read as words they also form a valid addition sum. Here is a simple example:

$$\begin{array}{r} T H R E E \\ T H R E E \\ T W O \\ T W O \\ + O N E \\ \hline E L E V E N \end{array}$$

This example is the smallest doubly-true English alphametic with a unique solution, where by smallest we mean it has the smallest sum word, 11.

87. Double-unique alphametic solutions are the most difficult to create and the rarest. The following few problems are my attempts at making a double-unique alphametic and to the best of my knowledge they have never been published:

```
          O N E
        N I N E
    T W E N T Y
  + F I F T Y
    E I G H T Y
```

88. Can you find the solution to my second double-unique alphametic?

```
        T H R E E
        S E V E N
          T E N
      T W E N T Y
    + T H I R T Y
      S E V E N T Y
```

89. These alphametic puzzles don't meet the text book definition of double-unique but I would suggest they are "1 ½ unique". True double-unique alphametics are just addends and sums, but in these cases I mixed it up a little:

```
        E  I  G  H  T
   +  T  W  E  L  V  E
            L  E  S  S
               N  I  N  E
   ─────────────────────
      E  L  E  V  E  N
```

90. See what I did here? 7 + 11 = 18 - 6 = 12

```
         S  E  V  E  N
   +  E  L  E  V  E  N
            L  E  S  S
                  S  I  X
   ─────────────────────
         T  W  E  L  V  E
```

91. This one is 8 + 11 = 19 - 7 = 12

```
         E  I  G  H  T
   +  E  L  E  V  E  N
            L  E  S  S
            S  E  V  E  N
   ─────────────────────
      T  W  E  L  V  E
```

92. 90 - 60 = 30

```
      N  I  N  E  T  Y
            L  E  S  S
   +  S  I  X  T  Y
   ─────────────────────
      T  H  I  R  T  Y
```

93. Here are a few silly ones just for fun. What is the number?

```
    F I V E
    F I V E
    N I N E
  + E L E V E N
    N U M B E R
```

94. 3 + Number + Number = 15, what are the numbers?

```
    T H R E E
    N U M B E R
  + N U M B E R
    F I F T E E N
```

95. Something + Something = 12, what's something?

```
    N U M B E R
  + N U M B E R
    T W E L V E
```

96. These are polite terms for the answer:

```
    H E F T Y
    H U S K Y
  + S T O U T
    O B E S E
```

97. Some may question the validity of this:

$$\begin{array}{r} \text{M E N} \\ + \text{W O M E N} \\ \hline \text{H A P P Y} \end{array}$$

98. But no one will question this:

$$\begin{array}{r} \text{A N D} \\ \text{O N E} \\ + \text{L O V E D} \\ \hline \text{H A P P Y} \end{array}$$

… and one plus loved equals happy.

99. This is another classic logic riddle: can you connect all 9 dots with only 4 straight lines? Hint: we are looking for a 2-dimensional solution, no spheres, cubes or wormholes, and no rearranging of the dots.

•   •   •

•   •   •

•   •   •

100. Form 4 equilateral triangles using only 6 toothpicks. Hint: you can't break the toothpicks!

Math Riddle Solutions

1. We'll call the first number *a* and the second number *b*. The difference between the two numbers can be represented as a - b, and the difference in their squares can be represented as $a^2 - b^2$, so:

$$a - b = 40$$
$$a^2 - b^2 = 4,800$$

The trick to these kinds of problems is to solve for a variable in one of the expressions and substitute the result into the other expression. Solving for *a* in the first expression we get a = b + 40, which we will substitute for *a* in the second expression:

$$(b + 40)^2 - b^2 = 4,800$$

We solve for $(b + 40)^2$ by using the FOIL method (First, Outside, Inside, Last):

$$(b + 40)(b + 40) = b^2 + 40b + 40b + 1,600$$

Which simplifies to:

$$b^2 + 80b + 1,600$$

So now our second expression can be expressed as

$$b^2 + 80b + 1,600 - b^2 = 4,800$$

$b^2 - b^2 = 0$ which eliminates b^2 from the expression:

$$80b + 1{,}600 - 1{,}600 = 4{,}800 - 1{,}600$$

Subtracting 1,600 from both sides leaves:

$$80b = 3{,}200$$

Dividing both sides by 80 to solve for b gives us:

$$b = 40$$

Now we can go back to our first equation and solve for a:

$$a - 40 = 40$$

Adding 40 to the left and right to solve for a gives us:

$$a = 80$$

Let's test our answers:

$$a - b = 40$$
$$80 - 40 = 40$$

$$a^2 - b^2 = 4{,}800$$
$$80^2 - 40^2 = 4{,}800$$
$$6{,}400 - 1{,}600 = 4{,}800$$

Both answers check good!

2. When we were asked for a number increased by its cube we can represent this as:

$$n + n^3 = 592{,}788.$$

To solve for *n* all we need to do is find the cube root of 592,788 and round to the nearest whole number:

$$\sqrt[3]{592{,}788} = 84.004$$

Rounding to the nearest whole number gives us 84.

Let's test our answer:

$$n + n^3 = 592{,}788$$
$$84 + 84^3 = 592{,}788$$
$$84 + 592{,}704 = 592{,}788$$

Our answer checks good!

3. We can represent this as

$$\sqrt[3]{n} = 10 \times \sqrt[4]{n} \text{ or}$$
$$n^{1/3} = 10n^{1/4}$$

The root of a number can also be expressed by raising the number to a fraction with the cube as the fraction's denominator.

To eliminate the two fractions ⅓ and ¼ we will need to raise both sides to the 12th power (this is the same as multiplying both sides by the denominators 3 and 4):

$$(n^{1/3})^{12} = (10n^{1/4})^{12}$$
$$n^{12/3} = 10^{12} \times n^{12/4}$$
$$n^4 = 10^{12} \times n^3$$

Dividing both sides by n^3 to solve for *n* gives us:

$$n^4 / n^3 = 10^{12}$$
$$n^4 / n^3 = n \quad \text{so}$$
$$n = 10^{12}$$

10^{12} is 1 with 12 zeros after it:

$$n = 1,000,000,000,000$$

Let's test our answer:

$$1,000,000,000,000^{1/3} = 10 \times 1,000,000,000,000^{1/4}$$
$$10,000 = 10,000$$

Our answer checks good!

4. The fence is square and 2 miles long so we know the perimeter p = 2 miles. The formula for the perimeter of a square is

$$p = 4s$$

Or 4 times the length of each side so now we know the length of each side of the fence is ½ mile:

$$p = 4s$$
$$p/4 = s$$
$$2/4 = s$$
$$s = \text{½ mile}$$

The area of a square is:

$$A = s^2 \text{ so}$$

$s = \frac{1}{2}^2 = 160$ acres for the 2-mile-long fence.

Solving for the 3-mile-long fence if P = 4s and P = 3 miles then each side of the fence is ¾ miles long. Now we can solve for the acreage of the area inside the 3-mile-long fence by using the ratio of the length of the fence to the acreage inside the fence:

$$\frac{\left(\frac{1}{2}\right)^2}{\left(\frac{3}{4}\right)^2} = \frac{160}{x}$$

The way we would read the previous expression would be "one-half squared is to 160 as three-quarters squared is to x". We can now represent the ratio as:

$$\frac{.25}{.5625} = \frac{160}{x}$$

$$.25x = 160 \times .5625$$

$$.25x = 90$$

$$x = 360 \text{ acres}$$

A 3-mile square fence will then enclose 360 acres.

5. This one is as easy as it looks: expressing the $90 you spent as a negative number since you spent it and the $100 as a positive number since you received it gives you:

$$-\$90 + \$100 = \$10$$

So, you made $10 from this transaction. Pretty obvious, but bear with me.

In the second transaction, you have $100 and spent $80 so we express this as:

$$\$100 + -\$80 = \$20$$

So, you are now $20 richer after these transactions. The important take away from this example is in the money world money coming in is positive and money going out is negative, which can also be expressed as ($x.xx), with the parentheses denoting the negative value. So, we could have expressed these transactions as:

$$(\$90) + \$100 = \$10$$
$$\$100 + (\$80) = \$20$$

6. To solve this problem we need to know the volume of a sphere:

$$4\pi \frac{r^3}{3}$$

Or 4 x Pi x the radius of the sphere cubed divided by 3.

If the Earth is 7,918 miles in diameter then its radius is 7,918/2 or 3,959 miles. The volume of the Earth then is:

$$V_{earth} = 4\pi \frac{3,959^3}{3} = 259,923,241,548 \text{ cubic miles.}$$

Since the Sun has a diameter of 865,000 miles and its radius is 865,000/2 or 432,500 miles its volume is:

$$V_{sun} = 4\pi \frac{432,500^3}{3} = 3.3888 \times 10^{17} \text{ cubic miles.}$$

The ratio of the volume of the Sun to the Earth can be expressed as:

$$V_{sun} / V_{earth} = 3.3888 \times 10^{17} / 259,923,241,548 = 1,303,773$$

Or a little over 1.3 million Earths will fit inside the Sun.

7. We can express the sum of the two numbers as a + b and their difference as a - b, so

$$a + b = 582$$
$$a - b = 218$$

Solving for *a* we get a = b + 218, and substituting our new value for *a* into the first equation:

$$(b + 218) + b = 582$$
$$2b + 218 - 218 = 582 - 218$$
$$2b = 364$$

Dividing both sides by 2 to solve for b:

$$2b / 2 = 364 / 2$$
$$b = 182$$

Now we can solve for a:

$$a + 182 - 182 = 582 - 182$$
$$a = 400$$

Let's test our answer:

$$400 + 182 = 582$$
$$400 - 182 = 218$$

The answer checks good!

8. A dozen writing pads and a box of pens both cost $17.18. We can express this as:

$$a + b = \$17.18$$

The difference in the value of the writing pads and the box of pens is $6.20. We can express this as:

$$a - b = \$6.20$$

Now we have enough information to solve for the values of a and b.

Solving for a in the second expression gives us:

$$a = b + 6.20$$

Substitute the value for a into the first expression:

$$(6.20 + b) + b = 17.18$$
$$6.20 - 6.20 + 2b = 17.18 - 6.20$$

Subtracting 6.20 from both sides leaves:

$$2b = 10.98$$
$$2b/2 = 10.98/2$$
$$b = 5.49$$

Solving for a then gives us:

$$a - 5.49 = 6.20$$
$$a = 11.69$$

So a dozen writing pads is $11.69 and a box of pens is $5.49.

9. We'll call Andrew a and Michael m. We can represent when Andrew gives Michael $10 and they both have the same amount of money as:

$$m + 10 = a - 10$$

We can represent when Michael gives Andrew $10 and Andrew has twice as much money as:

$$a + 10 = 2(m - 10)$$
$$a + 10 = 2m - 20$$

Solve for *a* in the first expression:

$$m + 10 + 10 = a - 10 + 10$$
$$m + 20 = a$$

Now substitute *a* in the second expression:

$$(m + 20) + 10 = 2m - 20$$
$$m + 30 = 2m - 20$$
$$m + 30 - m + 20 = 2m - 20 + 20 - m$$
$$m = 50$$

So Michael has $50. How much money does Andrew have?

$$m + 20 = a$$
$$a = 50 + 20 = 70$$

Andrew has $70.

Let's check our answer:

If Andrew gives Michael $10 he would then have:

$$\$70 - \$10 = 60$$

Michael would have:

$$\$50 + \$10 = \$60$$

The answer checks good!

10. Here's what we know:

$$ab = (a - b) + (a + b)$$
$$ab = a - b + a + b$$

There is a b on both sides so it is eliminated from the expression.

$$ab = 2a$$

Divide both sides by a to solve for b:

$$ab/a = 2a/a$$
$$b = 2$$

11. Twice the sum of two numbers can be expressed as:

$$2(a + b)$$

Twice the difference would then be:

$$2(a - b)$$

Twice the sum and twice the difference would be:

$$2(a + b) + 2(a - b) = 80$$
$$2a + 2b + 2a - 2b = 80$$

$2b - 2b$ eliminates b from the expression so:

$$4a = 80$$
$$a = 20$$

12. One half the sum of two numbers and one half the difference can be expressed as:

$$\tfrac{1}{2}(a + b)$$
$$\tfrac{1}{2}(a - b)$$

If one half the sum exceeds one half the difference by 60 then:

$$\tfrac{1}{2}(a + b) = \tfrac{1}{2}(a - b) + 60$$

This time we multiply both sides by 2 to get rid of the fractions:

$$2(\tfrac{1}{2}(a + b)) = 2(\tfrac{1}{2}(a - b)) + 2 \times 60$$
$$a + b = a - b + 120$$
$$a + b - a + b = a - b + 120 - a + b$$

There is an *a* on both sides so we can remove them leaving:

$$2b = 120$$
$$b = 60$$

13. If we let *h* equal Harry's age Mary's age can be represented as:

$$\tfrac{5}{6}h + 24 = \tfrac{11}{12}h$$

If we rewrite $\tfrac{5}{6}$ as $\tfrac{10}{12}$ we have 12 as the common denominator:

$$\tfrac{10}{12}h + 24 = \tfrac{11}{12}h$$

Multiply both sides by 12 to get rid of the fractions:

$$12(^{10}/_{12}h) + 12 \times 24 = 12(^{11}/_{12}h)$$
$$120h + 288 = 132h$$
$$120h + 288 - 120h = 132h - 120h$$
$$12h = 288$$
$$h = 24$$

So, when Harry and Sally married Harry was 24 and Mary was $^5/_6$ x 24 = 20 years old.

Let's check our answer: 24 years after their marriage Sally is $^{11}/_{12}$ her husband's age, so

$$^{11}/_{12} \times (24 + 24) = 44$$

If she was 20 years old when they married and this is 24 years later 20 + 24 = 44 so the answer checks good!

14. As fractions continue to get smaller and smaller their sum becomes closer to 1, so the answer technically would be 1 with an infinite number of decimal 9's (1.99999…), but for our purposes it is easier to just say the answer is "2".

15. A repeating decimal series is what we call a "repetend". An interesting thing about fractions with 9 as their denominator is their equivalent decimal value always results in a repetend: $^1/_9$ =

0.1111..., ²⁄₉ = 0.2222..., ³⁄₉ = 0.3333..., and you guessed it, our answer is ⁴⁄₉ since ⁴⁄₉ = 0.4444...

16. Since the density of ice is rounded to 0.9 we know 10% of the ice is sticking out of the water so the height of the cube is 24 inches (10 x 2.4 inches). Now it's just a matter of multiplying the height by the width and the length to get the volume:

$$2 \times 7 \times 8 = 112$$

The volume of the piece of ice is 112 cubic feet.

17. To find the three consecutive numbers find the cube root of 15,600 and round it to the nearest whole number less than the cube root and greater than the cube root and multiply them by the next whole number:

$$\sqrt[3]{15,600} = 24.9867$$

So, the numbers we are looking for are 24, 25, and 26.

Let's check our answer: 24 x 25 x 26 = 15,600 so the answer checks good!

18. This is just a tricky way of saying "find the number exactly between 1,042 and 1,236".

Subtract 1,236 from 1,042 to find the difference between the numbers and then half their value plus 1,042 is your answer:

$$1,236$$
$$-\ 1,042$$
$$194$$

$$194/2 = 97$$

$$1,042$$
$$+\ \ \ 97$$
$$1,139$$

1,139 is as much greater than 1,042 as it is less than 1,236.

19. To find the answer you need to take the cube root of 10,697, round down to the nearest whole number, find the cube of the number and subtract it from 10,697:

$$\sqrt[3]{10,697} = 22.0337$$

The nearest whole number is 22 so:

$$22^3 = 10,648$$

Find the difference between the two numbers:

$$10,697$$
$$-\ 10,648$$
$$49$$

So, 49 is the smallest number to be subtracted from 10,697 and make a perfect cube.

20. If we let *a* equal the number of bows one student makes and *s* equal the number of students then:

$$a = (s - 1) + 1 \quad \text{so}$$
$$a = s$$

Because the student can't bow to himself we subtract 1 and then add 1 to account for the instructor. We would then need to multiply *a* by the number of students in the class, *s*, which we now know is equal to *a*, so the number of bows can be represented as:

$$a \times a \text{ or}$$
$$a^2$$

If there are 900 bows total the number of students would be the square root of 900:

$$\sqrt{900} = 30$$

Since there are twice as many girls as boys in the class one third of the class must be boys:

$$\tfrac{1}{3} \times 30 = 10$$

There are 10 boys and 20 girls in the class.

21. To find the mathy girl's age we'll try the squares of single digit numbers starting with 3:

$$3^2 = 9$$

$\sqrt{9} + 3/8(9) = 3.04$, so we're way off, this number is too low.

$$4^2 = 16$$

$\sqrt{16} + 3/8(16) = 4 + (3 \times 16)/8 = 4 + 48/8 = 4 + 6 = 10$

This works! The mathy girl's age is 16.

22. If we let h represent the head, b represents the body, and t represent the tail this is what we know about the fish:

$$h = 9$$
$$t = h + \tfrac{1}{2}b$$
$$b = h + t$$

So, t = ½b + 9 since h = 9 and b = t + 9. Solving for b in the second equation:

$$b = 2(t - 9)$$

Now we know:

$$2(t - 9) = t + 9$$
$$2t - 18 + 18 - t = t + 9 + 18 - t$$
$$t = 27$$

If the tail is 27 inches then b = 36 inches, and

$$h + b + t = 9 + 36 + 27 = 72$$

The length of the fish is 72 inches or 6 feet.

23. If Mary is 24 years old and we let x = the difference between Mary and Ann's ages we can represent Ann's age as 24 – x.

"She is twice as old as Ann was when Mary was as old as Ann is now" sounds more confusing than it is. This is just a complicated way of saying Ann's age less the difference between their ages. Let's work through the phrase backwards, when Mary was as old as Ann is now would be represented as:

$$24 - x - x$$
$$24 - 2x$$

If Mary is twice as old as Ann was then could be represented as

$$24 = 2(24 - 2x)$$
$$24 + 4x - 24 = 48 - 4x + 4x - 24$$

24 -24 = 0 on the left, and 4x – 4x = 0 on the right, and 48 – 24 = 24 on the right, which leaves us with:

$$4x = 24$$

$$x = 6$$

The difference in Mary and Ann's ages is 6 years so Ann is 24 – x = 18 years old.

Let's check our answer:

$$24 = 2(24 - 2x)$$
$$24 = 2(24 - 2(6)) = 2(24 - 12) = 48 - 24$$
$$24 = 24$$

Our answer checks good!

24. The answer is those same 6 cats could eat 100 rats in 100 minutes, their collective rate of consumption is 1 rat per minute.

25. The boy filled the 3-quart container completely and emptied it into the 5-quart container. He did this again and ended up with one quart left over in the 3-quart container. Now all he should do is empty the 5-quart container, fill it with the quart from the other container and refill the 3-quart container. When he empties the 3-quart container into the 5-quart container he'll have exactly 4 quarts.

Another way for him to get exactly 4 quarts would be to fill the 5-quart container then fill the 3-quart container leaving 2 quarts in the 5-quart container. Empty the 3-quart container and pour what is left into the 3-quart container. Refill the 5-quart container, top off the 3-quart container and you will have exactly 4 quarts left over.

26. This is kind of a trick question; most people are quick to figure out the frog's rate of ascent is 1 foot per day and will predict it will take the frog 45 days to climb out of the well. The answer though is 43 days, on the 43rd day he only needs to climb 2 more feet to escape the well.

27. To get the whole family across the river first both sons cross since their combined weight is 150 pounds. One son returns the boat and lets his mother cross, and the other brother brings the boat back again. They then cross the river together again and one of the boys jumps out on the other side so the other boy can go back for his father. Once the father is across the boy on the other side goes back to pick up his brother for one last trip across.

28. The answer is neither, the fish displaces a volume of water equal to its weight leaving the aquarium 20 pounds.

29. We can represent the man's age as

$$(1/4 + 1/5 + 1/3)x + 13 = x$$

The common denominator is 60 which gives us

$$(15/60 + 12/60 + 20/60)x + 13 = x$$
$$47/60\, x + 13 = x$$

Multiply both sides by 60 to get rid of the fraction:

$$47x + 780 = 60x$$
$$780 = 13x$$
$$x = 60$$

The man was 60 years old.

30. We'll call Andrew a, Brandon b, and Connor c. This is what we know:

$$a = 80 + c$$
$$a = 1.2b$$
$$b = 1.25c$$

So, if $80 + c = 1.2b$ we can substitute the value of b for 1.25c:

$$80 + c = 1.2(1.25c) = 1.5c$$
$$80 = 1.5c - c = .5c$$
$$c = 160$$

If c is 160 then a is $160 + $80 = $240.

b is then 1.25 x 160 = 200.

Andrew is bringing $240, Brandon is bringing $200, and Connor is bringing $160.

31. Let a, b, and c be the boys' ages in order, h is the husband, w is the wife, and we'll use g for the grandfather.

$$h = w + b + 1$$
$$w = a + b + c + 3$$
$$a = b + c + 3$$
$$b = 4c + 1$$
$$g = h + w + a + b + c = 125$$

Substituting the value of *b* into the expression for *a*:

$$a = (4c + 1) + c + 3 = 5c + 4$$

Substitute these values into the expression for *w*:

$$w = (5c + 4) + (4c + 1) + c + 3$$
$$w = 10c + 8$$

Substitute what we know to solve for *h*:

$$h = (10c + 8) + (4c + 1) + 1$$
$$h = 14c + 10$$

Now we can solve for *g*:

$$g = (14c + 10) + (10c + 8) + (5c + 4) + (4c + 1) + c = 125$$
$$34c + 23 = 125$$
$$34c = 102$$
$$c = 3$$

The ages of the family members are:

The little brother is 3,

the middle brother is 13,

the older brother is 19,

the wife is 38,

and the husband is 52.

To check our answer the sum of the family's ages should equal the age of the grandfather, 125:

3 + 13 + 19 + 38 + 52 = 125, our answer checks good!

32. To calculate the size of the Texans ranch we solve the ratios of the number of cows to acres plowed and acres pastured:

$$3 \text{ cows} / 2{,}100 \text{ acres} = 700 \text{ acres plowed}$$
$$7 \text{ cows} / 2{,}100 \text{ acres} = 300 \text{ acres} \times 2 \text{ pastured} = 600 \text{ acres}$$
$$700 \text{ acres} + 600 \text{ acres} = 1{,}300 \text{ acres}$$

33. If g = value of gold and d = value of the dollar then

$$g = 1.25d$$

Solving for *d* we get

$$\frac{g}{1.25} = \frac{1.25d}{1.25} \qquad d = .80g$$

The value in gold of a dollar bill was $0.80

34. This problem is an arithmetic sequence, only in this case we know the sum of the series but not the amount each number in

the sequence increases by, or the total number of terms in the series. The formula for the sum of an arithmetic sequence is:

$$S = \frac{n(a_1+a_n)}{2}$$

where S is the sum, n is the number of terms, and a_1 and a_n are the first and last numbers in the series. Since we know the sum but not the number of terms n we multiply both sides by 2 and divide by the sum of the first and last numbers in the series:

$$n = \frac{2S}{a_1+a_n} \text{ so } \frac{2(1,120)}{(40+100)} = 16$$

The number of terms in the series is 16 which is also the number of days the trip will take.

The number of miles he increased his ride by each day, called the "common difference" can be found by

$$a_n = a_1 + (n-1)d$$

a_n is the last term, 100, a_1 is the first term, 40, n = the number of terms 16, now solve for *d*:

$$100 = 40 + (16-1)d$$
$$60 = 15d$$
$$d = 4$$

Cody rode 4 miles more each day for 16 days. The first day he rides 40 then 4 more each day until 100 miles:

40 + 44 + 48 + 52 + 56 + 60 + 64 + 68 + 72 + 76 + 78 + 84 + 88 + 92 + 96 + 100 = 1,120

This is 16 terms and the sum equals 1,120 miles so our answer checks good!

35. Even though we only know one side of the triangle we can still use Pythagorean's theorem since the height of the tree equals the hypotenuse of the triangle and the height of stump. Since the fallen over part of the tree forms the hypotenuse of the triangle we'll call it c, the distance from the base of the tree to the tip forms the base, b, so the stump is a. Since we don't know the height of the stump but we do know the height of the tree we'll say the hypotenuse is:

$$c = 120 - a$$

Plug this information into Pythagorean's Theorem and solve for a.

$$a^2 + b^2 = c^2$$
$$a^2 + 40^2 = (120 - a)^2$$

We'll use the FOIL method to solve for c^2:

$$c^2 = (120 - a)(120 - a)$$
$$c^2 = 14{,}400 - 120a - 120a + a^2$$
$$c^2 = 14{,}400 - 240a + a^2$$
$$a^2 + 1{,}600 = 14{,}400 - 240a + a^2$$

There's an a^2 on both sides which cancels out, leaving:

$$240a = 14{,}400 - 1{,}600$$
$$240a = 12{,}800$$
$$a = 53.33$$

If $a = 53.33$ then $c = 120 - 53.33 = 66.67$

The length of the fallen over part of the tree was 66.67 feet.

36. The log averages 40 miles per day, 80 miles during the day minus 40 miles at night. In 7 days, the log will have travelled 280 miles which means it will reach the mouth of the river on Friday.

37. For this problem we need to know the area of a sphere:

$$4\pi \frac{r^3}{3}$$

The volume of a 6-inch ball of yarn would be

$$4\pi \frac{3^3}{3} = 4\pi \frac{27}{3} = 4\pi \times 9 = 36\pi = 113.1$$

So, 113.1 cubic inches of yarn makes 1 pair of gloves.

The volume of a 12-inch ball of yarn would be

$$4\pi\frac{6^3}{3} = 4\pi\frac{6 \times 6 \times 6}{3} = 4\pi\frac{216}{3} =$$

$$4\pi \times 72 = 288\pi = 904.8$$

Did you see how you could solve most of the volume in your head? Don't let constants like π and exponents and fractions intimidate you! The volume of a 12-inch ball of yarn is 904.8 cubic inches. Now compare the ratio of volume of yarn to number of gloves and solve for the unknown:

$$\frac{113.1}{1} = \frac{904.8}{x}$$

$$113.1x = 904.8$$

$$x = 8$$

A 12-inch ball of yarn will make 8 pairs of gloves if a 6-inch ball makes 1 pair of gloves.

38. This riddle is much like the ball of yarn riddle. The volume of the 20-inch diameter melon is:

$$4\pi\frac{10^3}{3} = 4,188$$

The volume of the second melon is:

$$4\pi\frac{15^3}{3} = 14,137.2$$

Compare the ratios of the volume of melon to the price and solve for the unknown:

$$\frac{4{,}188.8}{2} = \frac{14{,}137.2}{x}$$

$$4{,}188.8x = 14{,}137.2 \times 2$$
$$4{,}188.8x = 28{,}274.4$$
$$x = 6.75$$

If the 20-inch melon sells for $2 the 30-inch melon should sell for $6.75.

39. To represent 2,234 as a thousandth multiply it by .001 or move the decimal point 3 places to the left:

$$\begin{array}{r} 2{,}234 \\ \times\ .001 \\ \hline 2.324 \end{array}$$

To represent 24,325 as a hundredth multiply it by .01 or move the decimal point 2 places to the left:

$$\begin{array}{r} 24{,}325 \\ \times\ \ .01 \\ \hline 243.25 \end{array}$$

Now it's just a matter of adding the two numbers together:

$$\begin{array}{r} 243.250 \\ +\ \ \ 2.324 \\ \hline 245.574 \end{array}$$

40. This riddle is like the previous problem. To represent 200 as hundredths we'll just move the decimal point two places to the left to get 2.00.

To represent 15 as tenths we move the decimal point 1 place to the left to get 1.50. Subtract the two numbers:

$$\begin{array}{r} 2.00 \\ -1.50 \\ \hline .50 \end{array}$$

41. The river problems are much like the dreaded train problems, once you learn how to do problems like this they will be a breeze. The key to these problems is to remember the relationship between distance, rate, and time:

$$\text{distance} = \text{rate} \times \text{time}$$

To solve for the rate of the current we let x = the rate of the boat in calm water and y = the rate of the current. Creating a table helps us keep track of what we know:

	distance	rate	time
upstream	24 miles	$x - y$	4 hours
downstream	24 miles	$x + y$	3 hours

Going upstream we're fighting the current so we subtract the rate of the current from the rate of the boat in calm water. Going downstream we're with the current so we add the rate of the current to the rate of the boat in calm water.

Since distance = rate x time we can express what we know as:

upstream: $24 = 4(x - y)$
$24 = 4x - 4y$
$6 = x - y$

downstream: $24 = 3(x + y)$
$24 = 3x + 3y$
$8 = x + y$

$6 = x - y$
$+8 = x + y$
$14 = 2x$
$x = 7$

The rate of the boat in calm water, x, is 7 miles per hour. Plug the value for x into the expressions for upstream and downstream to solve for y, the rate of the current:

upstream: $6 = x - y$
$y = x - 6$
$y = 7 - 6 = 1$

downstream: $8 = x + y$
$y = x - 8$
$y = 7 - 8 = -1$

The rate of the current is 1 mile per hour.

Another way to look at this problem would be to compare the distance they row in a certain amount of time to determine the rate of the current since distance = rate x time and distance / time = rate. Rowing downstream in 1 hour the crew can go 8 miles or ⅓ the total distance. Upstream they only row 6 miles which is ¼ the distance.

$$1/3 - 1/4 = 4/12 - 3/12 = 1/12$$

The difference between the distance they row downstream versus upstream in 1 hour is 1/12 or twice the difference the river flows in 1 hour so the rate of the current is 1 mile per hour.

42. First a refresher: the minuend is the number to be subtracted from and the subtrahend does the subtracting, the remainder is the difference:

minuend

<u>- subtrahend</u>

remainder

If we call the minuend m, the subtrahend s, and the remainder r we can represent what we know as:

$$m = 80 + s$$
$$s = 20 + r$$
$$r = m - s$$

We have *s* in all three expressions so we'll start with s = 20 + r and r = m - s.

Substituting m - s for *r* in the first expression gives us

$$s = 20 + (m - s)$$

We can't have *s* on both sides of the expression so we substitute m - 80 for *s* and work the equations from the inside out:

$$s = 20 + (m - (m - 80))$$
$$s = 20 + (m - m + 80)$$
$$s = 20 + m - m + 80$$
$$s = 100$$

If s = 100 then m = 80 + 100 = 180, and 180 - 100 = 80.

43. Since we are doubling the winnings and 32 shots were made we could represent this as

$$2^{32} = 4,294,967,295$$

Divide by 100 to convert from pennies to dollars or just move the decimal two places to the left for a grand total of $42,949,672.95!

If Logan had decided he wanted to be paid in pennies with each penny weighing 2.5 grams the total weight would have been 11,836 tons!

44. We know the rate of sound, we know the amount of time the sound traveled, and we know distance = rate x time.

1,125 feet/second x 10 seconds = 11,253 feet

A mile is 5,280 feet so the total distance in miles would be

11,253 feet / 5,280 feet/mile = 2.13 miles

We want to know how many miles and feet so we subtract two miles in feet from 11,253 to get the remainder:

11,253 feet - 2(5,280 feet/mile) = 693 feet

The thunder cloud is 2 miles and 693 feet away.

45. The answer is the products will be increased by the average of their sums, for example:

```
   13     13 + 1 = 14      14       238      13      14      29
 x 16     16 + 1 = 17    x 17     - 208    + 16    + 17    + 31
  208                     238       30      29      31      60
```

60/2 = 30

When 13 x 16 was increased by 1 the difference in the products was 30.

The average of the sums was also 30.

46. After the first bounce the ball has traveled 100 feet down + 50 feet up from the bounce for a total of 150 feet. Then it falls 50

feet and bounces 25 feet, falls 25 feet and bounces 12.5 feet, and so on. As the ball continues to bounce in this manner we get closer and closer to doubling the first drop and bounce of 150 feet for a total of *almost* 300 feet. If we consider this a thought experiment and assume the ball can bounce infinitely then it would continue to get closer and closer to 300 feet but never exactly reach it. In reality every time the ball transfers its potential energy developed by falling to kinetic energy from bouncing it will give up a little bit of energy each time in proportion to its coefficient of restitution, or its "bounciness".

47. Walking 4 miles per hour Sean can travel 1 mile in 15 minutes and at 5 miles per hour it will take him 12 minutes, so the difference in the two rates is

$$15 \text{ miles / hour} - 12 \text{ miles / hour} = 3 \text{ miles / hour}$$

The actual difference in time is he is either 10 minutes late or 20 minutes early, which is a total difference of 30 minutes.

If traveling 1 mile results in a difference of 3 minutes and the actual difference in time is 30 minutes then we can compare the ratios of distance per time and solve for the unknown distance:

$$\frac{1 \text{ mile}}{3 \text{ minutes}} = \frac{x}{30 \text{ minutes}}$$

$$3x = 30$$
$$x = 10$$

The total distance Sean must walk is 10 miles. Let's check our work:

If Sean walks 10 miles at 4 miles per hour he will reach his destination in:

$$\frac{4 \text{ miles}}{60 \text{ minutes}} = \frac{10 \text{ miles}}{x}$$
$$4x = 600$$
$$x = 150$$

So, walking 10 miles at 4 miles per hour will take him 150 minutes which is 10 minutes late, so he must need to be there in 140 minutes.

If Sean walks 10 miles at 5 miles per hour he will reach his destination in:

$$\frac{5 \text{ miles}}{60 \text{ minutes}} = \frac{10 \text{ miles}}{x}$$

$$5x = 600$$
$$x = 120$$

Walking 10 miles at 5 miles per hour will take him 120 minutes which is 20 minutes early, so he must need to be there in 140 minutes so our answer checks out!

48. Since the boat can go 20 miles downstream we need to know how long it would take the boat to go 20 miles upstream:

$$\frac{15 \text{ miles}}{1 \text{ hour}} = \frac{20 \text{ miles}}{x}$$

$$15x = 20$$

$$x = 20/15 = \tfrac{4}{3} \text{ or } 1\tfrac{1}{3}$$

Going upstream takes 1⅓ hours to go the same distance traveled downstream, so the difference is ⅓ of an hour greater or 20 minutes longer.

If we know the difference in time for how long it takes to travel part of the distance, and we know the total difference in time, we can compare the ratios of time per trip to solve for the total number of trips. 5 hours is 300 minutes so:

$$\frac{20 \text{ minutes}}{\text{trip}} = \frac{300 \text{ minutes}}{x}$$

$$20x = 300$$

$$x = 15$$

If the boat travels 20 miles per hour downstream and makes 15 trips then 20 miles x 15 = 300 miles.

49. We'll call Marliesa m and Jasmine j and make the following table:

	Now	Future
m	30	30 + t
j	3	3 + t

We can represent when m is 5 times greater than j as

$$m = 5j$$

We know at some point in the future

$$m = 30 + t$$

$$j = 3 + t$$

So, we could represent this as:

$$30 + t = 5(3 + t)$$
$$30 + t = 15 + 5t$$
$$30 + t - 15 - t = 15 + 5t - 15 - t$$
$$15 = 4t$$
$$t = {}^{15}/_4 \text{ or } 3¾ \text{ years}$$

Let's check our work:

In 3¾ years m = 30 + 3¾ = 33¾ and j = 3 + 3¾ = 6¾

6¾ x 5 = 33¾ so our answer checks good!

Here's an easy way to multiply 6¾ x 5 in your head:

Multiply 6 x 5 to get 30. 5 x ¾ is ¾ 5 times, and we know ¾ + ¾ = 1½, and 1½ and 1½ is 3 with a ¾ left over giving us an answer of 3¾. Adding 3¾ to 30 gives us the answer of 33¾.

50. In 1-hour David can row upstream ⅓ of the distance, and in 1 hour he can row downstream ½ the distance.

$$\tfrac{1}{3} - \tfrac{1}{2} = \tfrac{2}{6} - \tfrac{3}{6} = \tfrac{1}{6}$$

or twice the distance the river flows in 1 hour. So, the river flows $\tfrac{1}{12}d$ in 1 hour and since distance = rate x time:

$$\tfrac{1}{12}d = 1 \text{ mile/hour} \times 1 \text{ hour}$$
$$\tfrac{1}{12}d = 1 \text{ mile}$$

Multiply both sides by 12 to get rid of the fraction and we get 12.

David rowed 12 miles.

51. If we call the quarterback a and the linebacker b we can express the difference in the number of their steps as:

$$2b = 5a \text{ or}$$
$$a = \tfrac{5}{2}b$$

To find how many steps it takes for the linebacker to overcome the quarterback's 27 step lead we set the number of steps to x, subtract the linebacker's pace from the quarterback's and set that equal to 27:

$$\tfrac{5}{2}(5x) - 8x = 27$$
$$\tfrac{25}{2}x - 8x = 27$$

Multiply everything by 2 to get rid of the fraction.

$$2(^{25}/_2 x) - 2 \times 8x = 2 \times 27$$
$$25x - 16x = 54$$
$$9x = 54$$
$$x = 6$$

Remember x is the number of steps to overcome the quarterback. Now we need to multiply 6 by the number of the linebacker's steps, 5, to get $6 \times 5 = 30$.

The number of steps before the linebacker overcomes the quarterback's 27 step lead is 30.

52. We can represent the number of fish Morgan caught as:

$$^2/_5 x + 7 = 11$$

We multiply both sides by 5 to get rid of the fraction:

$$5(^2/_5 x) + 5 \times 7 = 5 \times 11$$
$$2x + 35 = 55$$
$$2x + 35 - 35 = 55 - 35$$
$$2x = 20$$
$$x = 10$$

Morgan caught 10 fish.

53. Did you get 5,000? Most people do, but the answer is actually 4,100! With the smaller numbers being in closer proximity to the larger number, 1,000, your brain wants to make

the answer bigger. The take away from this riddle is to remember sometimes your brain plays tricks on you!

54. If we let x = your age now, and y = my age now, then $y - x$ is the difference in our ages. We first need to find your age when I was your age now, which is a tricky way of saying your age less the difference between our ages:

$$x - (y - x)$$

Since I am now twice as old as you were then I can represent my age as 2 times the previous expression:

$$y = 2(x - (y - x))$$
$$y = 2(x - y + x)$$
$$y = 2x - 2y + 2x$$
$$y = 4x - 2y$$
$$3y = 4x$$
$$y = \tfrac{4}{3}x$$

If both our ages combined is 63 then we can say:

$$x + \tfrac{4}{3}x = 63$$

Multiply both sides by 3 to get rid of the fraction:

$$3x + 4x = 3 \times 63$$
$$7x = 189$$
$$x = 27$$

If $x = 27$ then $27 + y = 63$ and $y = 63 - 27 = 36$.

So, your age is 27 and my age is 36. To check our work when I was your age, 27, you were 27 - the difference in our ages (36 - 27) or 27 - 9 = 18. My age now, 36, is 2 x 18 so our answer checks out!

55. If we call the number of soldiers on each side n since she wants to form a square we can represent the number of soldiers as n^2. She had 39 soldiers left over so we can represent this as:

$$x = n^2 + 39$$

On her second try she increased n by 1 so this could be represented as:

$$x = (n + 1)^2 - 50$$

Solving for $(n + 1)^2$

$$(n + 1)(n + 1)$$
$$n^2 + n + n + 1$$
$$n^2 + 2n + 1$$

Adding this back into our expression $x = (n + 1)^2 - 50$ gives us:

$$x = n^2 + 2n - 49$$

Since the number of soldiers never changes we can use both expressions to solve for n, the number of soldiers on one side of the square.

$$n^2 + 39 = n^2 + 2n - 49$$

There is an n^2 on both sides of the equation so it drops out. Now we have:

$$39 = 2n - 49$$
$$39 + 49 = 2n - 49 + 49$$
$$88 = 2n$$
$$n = 44$$

Plugging the value for *n*, 44, into the first equation give us:

$$x = n^2 + 39$$
$$x = 44^2 + 39$$
$$x = 1{,}975$$

The officer has 1,975 soldiers in her formation.

56. When the tree fell over it formed a right triangle but we only know the length of the base which is 36 feet. We can call the hypotenuse the height of the tree less the side a or c = 96 - a. Plugging this into Pythagorean's Theorem gives us:

$$a^2 + b^2 = c^2$$
$$a^2 + 36^2 = (96 - a)^2$$

Now use the FOIL method to calculate c^2:

$$c^2 = (96 - a)(96 - a)$$
$$c^2 = 9{,}216 - 96a - 96a + a^2$$
$$c^2 = 9{,}216 - 192a + a^2$$

Plug this back into Pythagorean's Theorem:

$$a^2 + 1{,}296 = 9{,}216 - 192a + a^2$$

There is an a^2 on both sides which cancels out:

$$1{,}296 = 9{,}216 - 192a - 1{,}296 + 192a$$
$$192a = 7{,}920$$
$$a = 41.25$$

The height of the stump, a, is 41 feet 3 inches.

57. Cut the wood as shown and shift everything to the right.

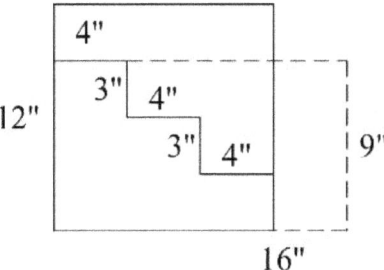

58. The first thing you should notice is $\frac{1}{3} + \frac{1}{4} + \frac{1}{5}$ does not equal a whole number. The least common denominator is 60 so the fractions are now

$$\tfrac{20}{60} + \tfrac{15}{60} + \tfrac{12}{60} = \tfrac{47}{60}$$

Zach gets $20, Dylan gets $15, and Cole gets $12, but $60 - $47 = $13 left over after the split. Now they need to find $\frac{1}{3}$, $\frac{1}{4}$, and $\frac{1}{5}$

of $13 and add those amounts back into what they've already calculated.

⅓ of 13 = 4.33 so we add $4.33 to Zach's $20 so he now has $24.33.

¼ of 13 = 3.25 so add $3.25 to Dylan's $15 so he now has $18.25.

⅕ of 13 = 2.6 so add $2.60 to Cole's $12 so he now has $14.60.

$4.33 + $3.25 + $2.60 = $10.18, and $13 - $10.18 = $2.82, so now they must divvy up $2.82. If you keep dividing the remaining amounts by ⅓, ¼, and ⅕ you will find Zach gets $25.53, Dylan gets $19.15, and Cole gets $15.32, and if you add these amounts together the sum is $60.

59. The answer is 8, if you came up with 112 you're thinking too hard! Picture each cat looking to its right and seeing the cat beside of it which gives you the 7 cats before each cat. A cat on every cat's tail is the cat behind each cat, giving you 8 cats total.

60. This is a classic logic problem. He first takes the goose across the river and drops it off. Then he goes back for the corn and drops it off on the other side and picks up the goose again and takes it back with him. He drops the goose off and takes the fox to the other side. Now he can go back for the goose and have them all safely across the river.

61. A 1-pound, 3-pound, 9-pound, and 27-pound weight will be all you need to weigh any amount from 1 to 40 pounds. The x and minus signs in the table on the next page show which sides of the scale the weights need to be on for each weight. For example, to weigh 34 pounds you would place a 27, 9, and 1-pound weight on one side and a 3-pound weight on the other, 27 + 9 + 1 = 37, and 37 - 3 = 34 pounds. You would place the item you want to weigh on the same side as the 3-pound weight. The table on the next page shows how this works for all weights between 1 and 40 pounds.

	1 pound	3 pound	9 pound	27 pound
1	x			
2	-	x		
3		x		
4	x	x		
5	-	-	x	
6		-	x	
7	-	x	-	
8	-		x	
9			x	
10	x		x	
11	-	x	x	
12		x	x	
13	x	x	x	
14	-	-	-	x
15		-	-	x
16	-	x	x	-
17	-		-	x
18			-	x
19	x		-	x
20	-	x	-	x
21		x	-	x
22	x	x	-	x
23	-	-		x
24		-		x
25	x	-		x
26	-			x
27				x
28	x			x
29	-	x		x
30		x		x
31	x	x		x
32	-	-	x	x
33		-	x	x
34	x	-	x	x
35	-		x	x
36			x	x
37	x		x	x
38	-	x	x	x
39		x	x	x
40	x	x	x	x

62. Technically, you will never run out of money because you are always dividing your money by 2 but after 19 days you'll have less than a penny and it will be hard to get change!

63. The answer is $78 + 15 + \sqrt{9} + \sqrt[3]{64} = 100$

64. Numbering the pennies clockwise from the upper right-hand corner you would place penny #5 on penny #4, penny #2 on penny #1, penny #11 on penny #10, and penny #8 on penny #7. Now there's 5 on each side!

65. The answer is:

$$35/70 + \sqrt{4}\,(\%) + 1^8 = \tfrac{1}{2} + 3 + 1 = 4\tfrac{1}{2}$$

66. Cut along the lines indicated in the drawing and shift the top piece to the right so the smaller sections are on top of each other and you'll have a 2' x 15' piece of wood.

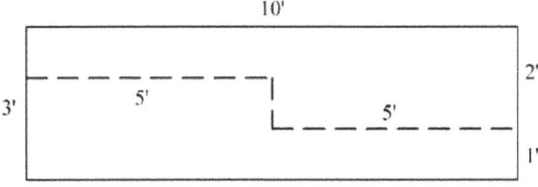

67. The two possible solutions are:

$$3^3 - 3 = 24$$

$$22 + 2 = 24$$

68. Neither. When you jump your motion is relative to the ship. However, if you jump off the ship...

69. There are 82 feet and 26 heads. You know horses have 4 hooves and people have 2 feet so you divide the number of heads by 2 and multiply one half by 4 and one half by 2 to see how far off you are:

$$26 \text{ heads} \div 2 = 13$$
$$13 \times 4 = 52$$
$$13 \times 2 = 26$$
$$52 + 26 = 78$$

So, you're off by 4. Let's try 14 horses and 12 riders:

$$14 \times 4 = 56$$
$$12 \times 2 = 24$$
$$56 + 24 = 80$$

You're getting closer, only off by 2! Now try 15 horses and 11 riders:

$$15 \times 4 = 60$$
$$11 \times 2 = 22$$
$$60 + 22 = 82$$

This is the right number of horses and riders, there are 15 horses and 11 riders.

70. There are many solutions, here are 3:

```
    1 5             5 6
    3 6               8
  + 4 7               4
    9 8         +     3
  +   2             7 1
  1 0 0         + 2 9
                1 0 0
```

```
    6 4
  + 2 5
    8 9  + 1 + 3 + 7 = 1 0 0
```

71. This is a trick question, when the conductors pass each other they will both be exactly the same distance from New York. Everything else is subterfuge!

72. The answers are shaded in gray:

```
            1   0   3   2   2   7   1
                            5   2   2   4
                6   3   9   9   7   4
                    8   8   9   4   7
                    3   0   5   4   1   7
      +     2   0   3   5   4   7   0
            4,  1   0   7,  3   0   3
```

73. The remaining numbers equal 1,111.

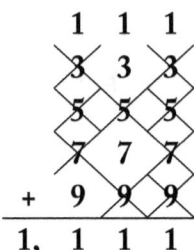

74. There are several ways to solve this. $24 \div 3 = 8$ so we need three 8 ounce portions. One way to split the juice three ways is to fill the 11 ounce and 5 ounce containers which leaves 8 ounces in the first container. Then pour from the 11-ounce bottle into the 13-ounce bottle, top it off with what's in the 5-ounce bottle, and pour what's left into the 11-ounce bottle. When you refill the 5 ounce bottle you'll have 8 ounces left.

75. The answer is never, if it continues to grow half as much as it did the day before the rate of growth gets slower and slower and approaches zero.

76. We can represent their present and past ages in the following table:

	Present	Past
d	71	- t
c	34	- t

We use - t because we're looking some number of years in the past. Their ages in the past can be represented as:

$$71 - t = 34 - t$$

If we want to know the point in the past when David was 3 times older than Chris we express this as

$$71 - t = 3(34 - t)$$
$$71 - t = 102 - 3t$$
$$71 - t + 3t - 71 = 102 - 3t - 71 + 3t$$
$$2t = 31$$
$$t = 15.5$$

When David was 71 - 15.5 = 55.5 years old Chris was 34 - 15.5 = 18.5 years old. 18.5 x 3 = 55.5 so our answer checks good!

77. We can represent their ages at some point in the future as $d = 71 + t$ and $c = 34 + t$.

	Present	Future
d	71	+ t
c	34	+ t

We can represent when Chris will be half his father's age as:

$$\tfrac{1}{2}(71 + t) = 34 + t$$

Multiply both sides by 2 to get rid of the fraction.

$$2(½(71 + t)) = 2(34 + t)$$
$$71 + t - 68 - t = 68 + 2t - 68 - t$$
$$3 = t$$

So, 3 years from now Chris will be half his father's age.

$$34 + 3 = 37$$
$$71 + 3 = 74$$
$$74 \div 2 = 37$$

Our answer checks good!

78. Any 2-digit number can be expressed as $10a + b$, so if we call the 2-digit number c then we can represent it as:

$$c = 10a + b$$

5 times the sum of the number's digits can then be expressed as:

$$c = 5(a + b)$$

Adding 9 to the number and reversing the digits can be represented as:

$$10a + b + 9 = 10b + a$$
$$10a + b + 9 - a - b - 9 = 10b + a - a - b - 9$$
$$9a = 9b - 9$$
$$a = b - 1$$
$$b = a + 1$$

Plugging our new value for b into the first equation gives us:

$$10a + b = 5(a + b)$$
$$10a + (a + 1) = 5(a + (a + 1)) = 5(2a + 1)$$
$$11a + 1 = 10a + 5$$
$$11a + 1 - 1 - 10a = 10a + 5 - 10a - 1$$
$$a = 4$$

Now we can solve for b:

$$10a + b = 5(a + b)$$
$$10(4) + b = 5(4 + b)$$
$$40 + b - b - 20 = 20 + 5b - 20 - b$$
$$20 = 4b$$
$$b = 5$$

It looks like our number is 45, let's try it out:

$$4 + 5 = 9, 5 \times 9 = 45$$
$$9 + 45 = 54$$

Our answer looks good!

79. There are many more solutions. The table on the next page has an example for each number:

1= 44÷44	11= 4÷.4 + 4÷4	21= (4.4+4)÷.4
2= 4÷4 + 4÷4	12= 4x4 - $\sqrt{4x4}$	22= (4+4)÷.4+$\sqrt{4}$
3= (4 + 4 + 4)÷4	13= 44÷4 + $\sqrt{4}$	23= 4÷(.4x.4)-$\sqrt{4}$
4= $\sqrt{(4 \times 4 \times 4)/4}$	14= 4÷.4+$\sqrt{4x4}$	24= (4+4)÷.4 + 4
5= $\sqrt{4x4}$ + 4 ÷ 4	15= 44÷4 + 4	25= (4+4+$\sqrt{4}$)÷.4
6= 4÷.4 - $\sqrt{4x4}$	16= 4x4 - 4 + 4	26= 4x4+4÷.4
7= 44÷4 - 4	17= 4x4 + 4÷4	27= 4÷(.4x.4)+$\sqrt{4}$
8= (4 + 4) x (4÷4)	18= 4÷.4 + 4 + 4	28= 44 - 4x4
9= 4 + 4 + 4÷4	19= (4+4 - .4)÷.4	29= 4÷(.4x.4) + 4
10= 4÷.4 + 4 - 4	20= 4÷.4 + 4÷.4	30= (4 + 4 + 4)÷.4

80. There are many solutions, here is one solution for each number.

1= 2^6÷(3+5)-7	11= 2x $\sqrt[3]{(3+5)}$+7	21= 2^4 + 3 +(-5 + 7)
2= 2^3 x 3!÷(-5^2+7^2)	12= 2^2 x 3!-(5+7)	22= 2^3 + ($\sqrt[3]{3+5}$ x 7)
3= -2 x $\sqrt[3]{(3+5)}$+7	13= 2^3 x 3!-5x7	23= -2^2 + 3 + (-5^2 + 7^2)
4= (23+5)÷7	14= $\sqrt{2^3 \text{x } 3!+(-5+7)^2}$	24= 2 x 3! + 5 + 7
5= 2^4-3-(-5+7)^3	15= 2^3+3^2-(-5+7)	25= 2^2 - 3 + (-5^2 + 7^2)
6= 2+3!-(-5+7)	16= 2^2 x 3!-(-5+7)^3	26= 2^2 x 3! + (-5 + 7)
7= 2^4+3-(5+7)	17= 2+3+5+7	27= 2^5 - 3 + 5 - 7
8= 2^5 x 3!÷(-5^2+7^2)	18= 2 x 3 + 5 + 7	28= 2^2 x 3! + (-5 + 7)^2
9= 2 x 3^2÷(-5+7)	19= 2^2 + 3 + 5 + 7	29= 2^2 + 3^2 + (-5 + 7)^4
10= 2 x 3!-(-5+7)	20= 2^2 x 3 +(-5 + 7)^3	30= 2^2 - 3! + 5^2 + 7

81. The 3 fractions in the father's will do not add up to 1. The first son got more than he should have, 9 instead of 8.5, the

second son got 6 instead of 5.67, and the third son got 2 instead of 1.89. There should have been 17/18 of a cow left over (.944).

82. Make the 4 cuts as shown below:

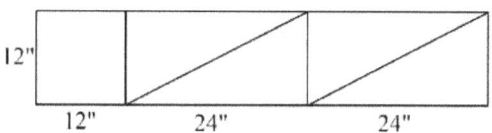

And rearrange them to look like this:

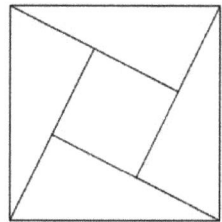

If the base of the triangle is 24 inches and the height is 12 inches than $a^2 + b^2 = c^2$ is $12^2 + 24^2 = \sim 26.83$ inches, which is close enough to 27 inches for our purposes.

83. A few possible solutions for this riddle are:

```
    7 5 2        4 3 7        3 6 5
+   3 4 6    +   5 8 9    +   7 2 4
---------    ---------    ---------
 1, 0 9 8     1, 0 2 6     1, 0 8 9
```

There are 93 more possible solutions!

Here are a few tips on how to solve alphametic riddles. If C + F = J, B + E = I, and A + D = GH, then A + D must be 10 or greater.

No combination of numbers from 0-9 results in a number greater than 17 so G must be 1. Starting from left to right and working with the first column try different combinations of numbers which will have sums of 10 or more, including sums of 9 like 2 + 7, 3 + 6, and 4 + 5 in case you carry a 1. Keep track of the numbers you use on the side of the page, I like to make a row of numbers from 0 - 9 and strike out the numbers used. Then try different combinations of the remaining numbers in the other two columns. With a little logic and patience you should start finding many working combinations of numbers.

84. There is only one solution for this classic H. E. Dudeney alphametic riddle:

$$\begin{array}{r} 9,567 \\ +1,085 \\ \hline 10,652 \end{array}$$

D=7, E=5, M=1, N=6, O=0, R=8, S=9, and Y=2.

85. The solution to this alphametic is:

$$\begin{array}{r} 698,392 \\ +3,192 \\ \hline 701,584 \end{array}$$

A=5, B=7, E=9, I=0, N=1, O=3, R=8, S=2, Y=4, and Z=6.

86. The answer to our first double-unique alphametic riddle:

```
      8 4, 6 1 1
      8 4, 6 1 1
          8 0 3
          8 0 3
  +       3 9 1
  ─────────────
  1 7 1, 2 1 9
```

E=1, H=4, L=7, N=9, O=3, R=6, T=8, V=2, and W=0.

87. The solution to my first double-unique alphametic is:

```
          9 8 4
        8, 5 8 4
    3 6 4, 8 3 2
  + 7 5, 7 3 2
  ─────────────
    4 5 0, 1 3 2
```

E=4, F=7, G=0, H=1, I=5, N=8, O=9, T=3, W=6, and Y=2.

88. The solution to my second double-unique alphametic is:

```
        8 6, 5 7 7
        1 7, 9 7 0
            8 7 0
        8 2 7, 0 8 3
  +     8 6 4, 5 8 3
  ───────────────────
    1, 7 9 7, 0 8 3
```

E=7, H=6, I=4, N=0, R=5, S=1, T=8, V=9, W=2, and Y=3.

89. The answer to our first not quite double-unique alphametic:

```
        8 1, 6 0 7
    + 7 4 8, 3 9 8
             3, 8 2 2
             5, 1 5 8
    ─────────────────
        8 3 8, 9 8 5
```

E=8, G=6, H=0, I=1, L=3, N=5, S=2, T=7, V=9, and W=4.

90. Seven + Eleven - Six = Twelve:

```
        4 1, 6 1 0
    + 1 8 1, 6 1 0
             8, 1 4 4
                4 9 7
    ─────────────────
        2 3 1, 8 6 1
```

E=1, I=9, L=8, N=0, S=4, T=2, V=6, W=3, and X=7.

91. Eight + Eleven - Seven = Twelve:

```
        2 1, 6 4 3
    + 2 5 2, 8 2 0
             5, 2 9 9
             9 2, 8 2 0
    ─────────────────
        3 7 2, 5 8 2
```

E=2, G=6, H=4, I=1, L=5, N=0, S=9, T=3, V=8, and W=7.

92. Ninety - Sixty = Thirty:

```
      1 8 1, 0 2 3
             9, 0 7 7
    + 7 8, 4 2 3
    ─────────────────
      2 6 8, 5 2 3
```

E=0, H=6, I=8, L=9, N=1, R=5, S=7, T=2, X=4, and Y=3.

93. Five + Five + Nine + Eleven = Number

```
        9, 7 4 1
        9, 7 4 1
        2, 7 2 1
   +  1 8 1, 4 1 2
   ─────────────────
      2 0 3, 6 1 5
```

B=6, E=1, F=9, I=7, L=8, M=3, N=2, R=5, U=0, and V=4.

94. Three + Number + Number = Fifteen:

```
        9, 7 4 1
        9, 7 4 1
        2, 7 2 1
   +  1 8 1, 4 1 2
   ─────────────────
      2 0 3, 6 1 5
```

B=3, E=0, F=1, H=6, I=7, M=9, N=8, R=4, T=5, and U=2.

95. Number + Number = Twelve:

```
       4 7 8, 6 0 3
   +   4 7 8, 6 0 3
   ─────────────────
       9 5 6, 1 2 6
```

B=0, E=6, L=1, M=8, N=4, R=3, T=9, U=7, V=2, and W=5.

96. Hefty + Husky + Stout = Obese:

$$
\begin{array}{r}
2\,1,753 \\
2\,8,403 \\
+\ 4\,5,985 \\
\hline
9\,6,141
\end{array}
$$

B=6, E=1, F=7, H=2, K=0, O=9, S=4, T=5, U=8, and Y=3.

97. Men + Women = Happy:

$$
\begin{array}{r}
831 \\
+\ 4\,9,831 \\
\hline
5\,0,662
\end{array}
$$

A=0, E=3, H=5, M=8, N=1, O=9, P=6, W=4, and Y=2.

98. And One + Loved = Happy:

$$
\begin{array}{r}
164 \\
967 \\
+\ 2\,9,874 \\
\hline
3\,1,005
\end{array}
$$

A=1, D=4, E=7, H=3, L=2, N=6, O=9, P=0, V=8, and Y=5.

99. First, draw a diagonal line from the lower right corner to the upper left. Then draw a line from the upper left-hand corner to below the bottom left hand corner (I never said you had to work inside the dots!). Now draw a line through the bottom middle dot and the middle dot on the right side and continue the line until it's even with the top row. Then all you have to do is draw a line to the first dot and you're done!

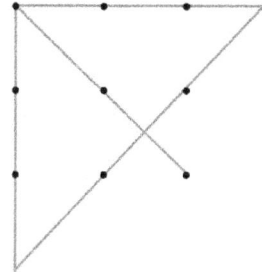

100. The trick is to think 3-dimensionally and make a tetrahedron:

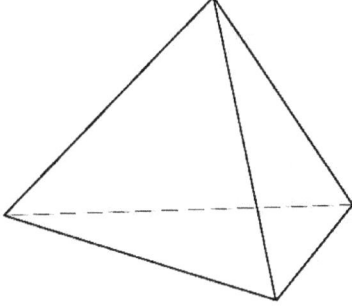

The 6 toothpicks now make 4 equilateral triangles.

Answer to the geometric riddle: The officer arranged his men in the shape of a dodecagon, a 12-sided polygon, and stood in the middle. He has 120 men but by arranging them in a dodecagon it seems like he has 12 x 11 men, 132.

Solution to the Pop Quiz: Neither! 8 + 5 = 13.

Answer to the 1st Math Funny: Division.

Answer to the 2nd Math Funny: Because it was less than 32 degrees.

Answer to the 3rd Math Funny: i<3u

Answer to the 4th Math Funny: To get to the same side!

Answer to the 5th Math Funny: Because 7 ate 9!

Answer to the 6th Math Funny: Algebros.

Answer to the 7th Math Funny: A roamin' numeral.

Answer to the 8th Math Funny: Big hands.

Answer to the 9th Math Funny: A belly ache.

About the Author

David is an educator, engineer, and veteran with a passion for getting kids interested in science, technology, engineering, and mathematics. He has worked with over 5,000 youths while running a Department of Defence STEM program, sharing with them the fun and excitement of STEM activities. He also has over 20 years of experience working for the federal government, refineries, and chemical plants as a mechanical engineer. In 2017 he started KidsSci Inc. so he could go back to sharing with kids the awesomeness of science and math.

Crush Your Math Fear! is his first book in a series on STEM activities. He chose to tackle math first because he thinks the M in STEM should come first but then you would have to call it METS, which we think is already trademarked, or MEST, which the marketing people probably wouldn't approve.

Even though little David was "one of those smart kids" (not a cool thing back in the 70's) he struggled with math at a young age. This is the book he wished he had when he was young. He hopes it will find its way into the hands of kids like he was as well teachers and parents interested in helping kids with math but not finding the right resources.

To connect with David and KidsSci go to:

https://KidsSci.com
https://facebook.com/KidsSci
https://twitter.com/KidsSci

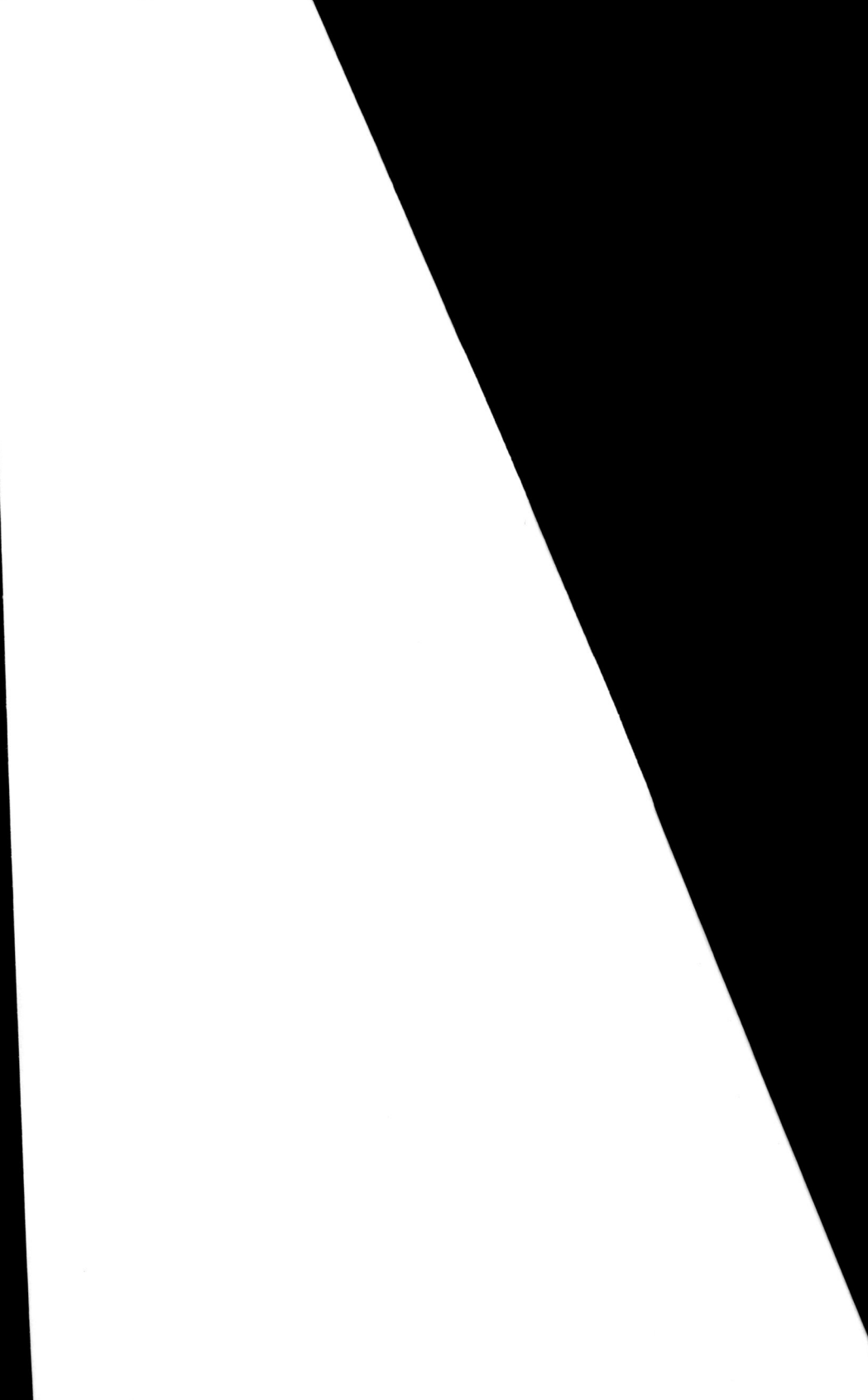

www.ingramcontent.com/pod-product-compliance
Lightning Source LLC
Chambersburg PA
CBHW071451040426
42444CB00008B/1294